PLANTING
THE FUTURE

PLANTING
THE FUTURE

Developing an Agriculture that Sustains Land and Community

Edited by
Elizabeth Ann R. Bird, Ph.D.
Gordon L. Bultena, Ph.D.
John C. Gardner, Ph.D.

**Research and Publication
Underwritten by**
Northwest Area Foundation
St. Paul, Minnesota

Publication Coordinated by
Center for Rural Affairs
Walthill, Nebraska

 Iowa State University Press / Ames

This book has been produced directly from camera-ready copy supplied by the editors.

No part of this book may be reproduced in any form or by any electronic or mechanical means, including information storage and retrieval systems, without permission in writing from the copyright holder, except for brief passages quoted in review.

∞ Printed on acid-free paper in the United States of America

First edition, 1995
Second printing, 1985

Library of Congress Cataloging-in-Publication Data

Planting the future: developing an agriculture that sustains land and community/edited by Elizabeth Ann R. Bird, Gordon L. Bultena, John C. Gardner.—1st ed.
        p.   cm.
    Includes bibliographical references (p.   ) and index.
    ISBN 0-8138-2072-3
    1. Sustainable agriculture—United States. 2. Agriculture—United States. 3. Sustainable agriculture—Research—United States. 4. Agriculture—Research—United States. I. Bird, Elizabeth Ann R. II. Bultena, Gordon L. III. Gardner, John C.
S441.P58   1995
338.1'0973—dc20

                                                                    94-41959

CONTENTS

CONTENTS

CONTENTS

CONTENTS

Sustainable agriculture (also called alternative agriculture) is controversial. Claimed by some to be environmentally sound, profitable, and supportive of America's family farms, it is condemned by others as impractical and unprofitable—and, why change conventional agriculture, a proven performer?

Little comparative work has been done on sustainable and conventional agriculture, so little is known about the relative benefit and cost of each farming style. Advocates for each often are polarized into opposing camps, sharing similar concerns but disagreeing strongly about how to proceed.

These facts led the Northwest Area Foundation to sponsor a systematic socioeconomic comparison of sustainable and conventional farming to inform taxpayers, policymakers, consumers, farmers, ranchers, environmentalists, farm groups, and others concerned about rural development. This comparison formed the centerpiece of the Foundation's **Sustainable Agriculture Initiative,** a five-state, multi-institutional cluster of research grants.

In an innovative approach, grants were made to joint teams of alternative agriculture organizations and academic institutions in Iowa, Minnesota, North Dakota, Montana and Oregon. Funding included support for research and demonstration initiated at the state level.

This book distills the results of the Initiative. It explains the methodology and findings from socioeconomic studies conducted across four of the states. It also synthesizes lessons learned from the farm-level research and demonstration projects, and from the Initiative as a whole. Importantly, this work concludes with an agenda for future studies, and public policy implications of the reported research.

The Foundation also supported a number of efforts independent of the Initiative, including demonstration, education, and research across the Foundation's eight-state region (Minnesota, Iowa, the Dakotas, Montana, Idaho, Washington, and Oregon). Some projects actually began prior to the Initiative; others attempted to fill critical gaps that became apparent after the Initiative began.

Because the Foundation's efforts were a reaction to the dearth of empirical information on how specific farm policy options affect farm economies and the relative profitability of sustainable agriculture, three projects were designed to provide such analysis. Another project focused on the dynamics of farm community change in relation to sustainable agriculture. Findings from a number of these related studies are highlighted in sidebars throughout this book.

As this book amply demonstrates, analysis of sustainable agriculture's potential has been started but not completed. Many important issues could not be addressed within the financial, personnel, and time resources available. Regardless, this multidisciplinary, multi-institutional, multistate effort is an important first step.

For the first time, people from various locales, disciplinary perspectives, and beliefs were brought together to systematically compare conventional and sustainable agricultural systems on economic, social, and ecological criteria at the farm,

farm-family, and community levels. This process of coming together has been an accomplishment in itself. It not only generated the findings reported here, but forged valuable human relationships as well.

For the policymaker, administrator, farmer, scientist, or consumer, the following pages offer three things: (1) research-based analysis on the socioeconomic impacts of sustainable agriculture, (2) agricultural research alternatives, and (3) policy analysis. This book provides new, high-quality information to inform the public dialogue on future directions for agriculture and public policy.

This study is a major step forward in comparing the impacts of sustainable and conventional farming practices on farms, farm families, and agriculture-dependent communities from a variety of intellectual and organizational perspectives. It represents a new chapter in the debate about linkages among farm policy, rural vitality, environmental quality, and agriculture. The Foundation hopes it is a chapter that will spur research and action toward developing an agriculture that sustains land and community.

Small Community and Rural Development   Karl N. Stauber*
U.S. Department of Agriculture   Deputy Undersecretary
Washington, DC   1 January 1995

*Formerly Vice President/Program, Northwest Area Foundation.

With only two percent of the U.S. population farming the land, few Americans give a second thought to the source of their food and fiber. Fewer still pause to consider how agricultural practices affect the natural environment and rural communities. Yet, as a nation, we face critical choices today that will determine the future character of our rural landscape.

- Will farming practices be designed to protect and nurture the environment, for example by providing wildlife habitat, building soil quality, and protecting crop genetic diversity? Will they be designed to conserve limited natural resources such as oil, minerals, and groundwater? Or will farming react to environmental concerns in a piecemeal fashion, and only under governmental pressure?

- Will young people find ample opportunity in self-employed, full-time farming? Or will farm ownership and control be accessible only to a few well-capitalized families and corporations?

- Will stable or growing farm numbers and a robust farming population help arrest the decline of rural communities and contribute to their revitalization? Or will agriculture-dependent rural communities continue to stagnate and disappear as rural populations and economic opportunity dwindle?

- Will farms retain more value from food and fiber production to contribute wealth to rural communities? Or will the value be captured by large, urban-centered agribusinesses?

Planting the Future provides information to help this nation respond thoughtfully to the preceding questions. In particular, this book examines:

- Which approaches to farming hold the most promise for simultaneously protecting the natural environment, maintaining farm entry opportunities, and revitalizing rural communities.

- What public action is needed to enable and encourage farmers to use these methods.

Planting the Future reports key findings of a multi-state research initiative sponsored by the Northwest Area Foundation. Coordinated research in four states—Iowa, Minnesota, Montana, and North Dakota—identified farmers who take two distinctly different approaches to agricultural production. The study obtained information with which to explore the impact of each alternative on the future structure of agriculture and the character of rural communities.

We term these two agricultural approaches "conventional" and "sustainable." For purposes of this study, *conventional farmers* are those with highly specialized cropping systems that rely on substantial amounts of synthetic crop nutrients and pesticides. *Sustainable farmers* limit use of synthetic fertilizers or pesticides, employ farm-produced resources and cultural management, raise more diverse crops and livestock, and express commitment to environmental sustainability (see

Chapter 3).

We found that the personal characteristics and farm structural characteristics of sustainable farmers are more conducive to maintaining vital rural communities and environmental quality. Here are some interesting findings from the research:

- Sustainable farmers plant less of their land in row crops like corn and soybeans and more in soil-conserving small grains, hay, forage, and pasture. They also keep more land in woodlands, wetlands, and other conservation uses (Chapter 4).

- Sustainable farms typically are smaller than conventional farms in three of the four states studied, although farm size varies widely within both farm types (Chapter 4).

- Sustainable farmers work about one-fourth more hours than conventional farmers. Both farm operators and their spouses work more hours on sustainable farms. But, since much of the work on sustainable farms pertains to livestock, it is spread more evenly through the year (Chapter 5). Labor and management demands on sustainable farms are greater and may constrain farm size.

- Sustainable farmers hire proportionately more nonhousehold labor than conventional farmers. Yet, nonhousehold workers on sustainable farms are more likely to have an ownership interest in the operation (Chapter 5).

- The best-performing sustainable farms are often smaller, and often make greater use of crop rotations and on-farm sources of crop nutrients (Sidebar 6-4).

- Although sustainable farmers spend less overall on production inputs than conventional farmers, they more often purchase goods and services produced locally by both farmers and nonfarmers, and they spend more per acre on these goods and services (Chapter 7).

- Sustainable farmers are as likely as conventional farmers to participate in and provide leadership to farm, church, and civic groups (Chapter 8).

- Sustainable farmers are more concerned about environmental problems than are conventional farmers (Chapter 8). Their approach to farming often is motivated more by environmental and health factors than by economic considerations (Chapter 9).

- Sustainable farmers are more concerned about the uncertain future of their local communities than are conventional farmers (Chapter 8).

- Sustainable farmers are more confident than conventional farmers that their children or other family members will continue to farm (Chapter 4).

These and other findings suggest that a widespread shift to sustainable farming practices would mean a greater number of farms, stability or growth in the rural population, increased farm employment, more value-adding economic activity on farms and in rural communities, and a farm population that is more committed to securing a sustainable future for farming and rural communities.

However, the results also suggest that a widespread shift to sustainable farming is unlikely, unless changes are made in federal commodity and conservation programs, rural development policy, and publicly supported research and extension education. For example, the research disclosed that:

- In 1991, the year these data were collected, sustainable farms generally reported lower net farm income and lower returns on equity (Chapter 6). However, the variation within each group was great; consequently, little of the difference may be directly attributable to their use of "sustainable" versus "conventional" practices (Sidebar 6-3). In general, the evidence on the economic performance of sustainable versus conventional farms is mixed (Sidebar 6-1).

- Some sustainable farmers are doing very well (see Sidebars 6-5 and 6-6). The top one-third of the sustainable farms surveyed performed competitively with the conventional farmers (Sidebar 6-4).

- Wide variability in the success of sustainable farmers suggests that some are finding the information and support they need, but others are not. Organized support may be critical to greater success (Sidebar 6-2). Yet, sustainable farmers seemingly are not receiving adequate support from public research and educational institutions. Sustainable farmers more often obtain useful information from their own experiments and from fellow sustainable farmers than they do from extension agents, university researchers, or government agency personnel (Chapter 9).

- The concerns of sustainable and conventional farmers differ. For example, sustainable farmers report greater stress about timely completion of farm tasks and complexity in decision making (Chapter 8).

- Availability of information about alternative agricultural practices was a top concern of sustainable farmers in Montana, North Dakota, and Iowa (Chapter 9).

These findings are important to the future direction of agricultural science and education. The research and extension needs of sustainable farmers differ from those of conventional agriculture. Sustainable agriculture requires long-term, whole-farm analysis, and it benefits from active participation of farmers in the research.

Current public research and extension programs need to provide leadership in creating a more sustainable agriculture. *Planting the Future* describes important benefits to sustainable farming systems from some alternative approaches to agricultural research and extension. Sustaining agriculture for the long term will require more systems-oriented study at all levels. Promising methods are:

- *Participatory research involving farmers* (Chapter 10).
- *Case studies and decision cases* (Chapter 11).
- *Use of ecological concepts and theory* to integrate the study of agriculture

(Chapter 12).

- *Use of geographical information systems* (Chapter 13).

Areas of research especially needed for sustainable farming include whole-farm management systems and problem-solving approaches that reduce input costs and add value to farms (Sidebar 15-1).

Changes in public research policy can promote the development of information vital to sustainable farming's success. These changes include strengthening the U.S. Department of Agriculture's programs for sustainable agriculture research, redirecting public funds to information and management-intensive technologies that directly benefit sustainable farmers, supporting farmer participation in research, targeting extension efforts to the development of farmers' decision-making skills, and focusing genetic research and new crop utilization research on the diverse crops produced on sustainable farms (Chapter 16).

Findings point also to needed changes in federal commodity programs:

- Federal commodity programs are biased against sustainable farming. The greater diversity of crop rotations used by sustainable farmers means they have less acreage planted in crops that are eligible for deficiency payments. This reduces the profitability of sustainable farming, discourages conversion to sustainable practices, and places sustainable farmers at a disadvantage in the land and commodity markets (Chapter 16).

Commodity programs could be redesigned to remove these penalties on diversified sustainable farming systems. Farm program payments could be targeted to protect farmers who have smaller acreages planted in program crops. Rather than paying farmers not to farm "set-aside" acres, an environmental reserve could pay farmers to reduce production in environmentally beneficial ways (Chapter 16).

Rural development policies also are critical to promoting sustainable farming:

- Sustainable farmers seem to have fewer opportunities to buy needed inputs and to sell their products close to home (Chapter 7). They are creating a demand, largely unfilled in their local communities, for new rural business ventures, such as processing plants and marketing services for the alternative crops they produce.

- Sustainable farmers produce a greater variety of commodities on a smaller scale (Chapter 4). This means that they face a disadvantage in marketing, because they forego volume bonuses.

Key rural development strategies for farm-based communities are defining markets that support sustainable farming, finding rural entrepreneurs to take advantage of these opportunities, and supporting their business development efforts. Rural development policy could support such strategies. Policies that enforce antitrust legislation and that support producer cooperatives could help smaller-volume producers overcome their market disadvantages (Chapter 16).

Results of the four-state study provide some answers, but raise many questions. Further research is needed to answer these questions (Chapter 15):

- Why are some sustainable farmers so much more successful than others?
- Why are sustainable practices correlated with smaller farm size?
- What are the long-term social and environmental benefits of farm diversification?
- How can sustainable farmers' unmet needs best be accommodated?
- How might a broader shift to sustainable farming affect the structure of rural economies?
- How could public agricultural research and extension programs be redirected to better support sustainable farming?

Though further research is needed, this unique and timely book provides research-based information to answer the biggest question: *could a shift to more sustainable agricultural practices foster the multiple societal goals of environmental protection, farm-based economic opportunity, and vital rural communities?* Answering in the affirmative, this work offers a strategy for promoting this shift through alternative research and extension approaches and through changes in federal policy.

ACKNOWLEDGEMENTS

We gratefully thank participants from these alternative agriculture groups and universities, many of whom worked in teams, as indicated:

Alternative Energy Resources Organization/Montana State University
The Center for Rural Affairs
Iowa State University/Practical Farmers of Iowa
Land Stewardship Project/University of Minnesota
University of Minnesota Cartography Laboratory
North Dakota State University/Northern Plains Sustainable Agriculture Society
Oregon Tilth/Oregon State University
South Dakota State University
Tufts University School of Nutrition
Virginia Polytechnic Institute and State University
Washington State University
University of Wisconsin/LaCrosse

We also wish to express our gratitude to the many farmers, sustainable and conventional, who took time to respond to our mail questionnaires, telephone interviews, and personal interviews, and who in many cases agreed to have allow scientists to collect more in-depth production and economic data at their farms.

Many people at the Northwest Area Foundation played critical roles in this endeavor. They include members of the Board of Directors and Karl Stauber, Marty Strange, Terry Saario, Vicki Itzkowitz, Victoria Tirrel, and Kevin Thompson. Also vital to the success of this endeavor have been individuals at the Center for Rural Affairs, particularly Elizabeth Bird, who functioned as coordinator and managing editor for this volume, and Chuck Hassebrook. Special thanks also to science writer Fred Schroyer, who edited this work.

C ontributors to this work are acknowledged in each chapter and sidebar. Alphabetically, they are:

Elizabeth Ann R. Bird, Center for Rural Affairs, Research and Technology Policy Project Leader

Alfred Merrill Blackmer, Iowa State University, Professor of Agronomy

Gordon L. Bultena, Iowa State University, Professor of Sociology

Charlene Chan-Muehlbauer, Minnesota Department of Agriculture, Energy and Sustainable Agriculture Program Research Analyst

Gregory Chu, University of Wisconsin/LaCrosse, Professor of Geography and Earth Science

Sharon A. Clancy, North Dakota State University, formerly Research Associate, Carrington Research and Extension Center

Sheila M. Cordray, Oregon State University, Associate Professor of Sociology

Jodi Dansingburg, Land Stewardship Project, Program Organizer/Policy Analyst

Richard P. Dick, Oregon State University, Associate Professor of Crop and Soil Science

Thomas L. Dobbs, South Dakota State University, Professor of Economics

John W. Duffield, University of Montana, Professor of Economics

Derrick N. Exner, Iowa State University, ISU Extension/Practical Farmers of Iowa Farming Systems Coordinator

Cornelia Butler Flora, Iowa State University, North Central Regional Center for Rural Development, Director and Professor of Sociology; formerly Virginia Polytechnic Institute and State University, Professor and Head of Sociology Department

John C. Gardner, North Dakota State University, Carrington Research and Extension Center, Superintendent/Agronomist

Gary A. Goreham, North Dakota State University, Assistant Professor of Sociology

Douglas Gunnink, formerly Minnesota Department of Agriculture, Energy and Sustainable Agriculture Program, On-Farm Research Coordinator

Charles (Chuck) Henry Hassebrook, Center for Rural Affairs, Leader—Stewardship and Technology Program

Eric O. Hoiberg, Iowa State University, Professor of Sociology

Keith Jamtgaard, formerly Montana State University, Assistant Professor of Sociology

Susan Kathryn Jarnagin, Iowa State University, Research Assistant in Sociology

Brenda J. Johnson, Cornell University Department of Sociology, Ph.D. candidate, and former intern, Center for Rural Affairs

Frederick Kirschenmann, Kirschenmann Family Farms, Manager

Ron Kroese, National Center for Appropriate Technology, President; formerly Land Stewardship Project, Executive Director

CONTRIBUTORS

Al Kurki, Consultant to the Northwest Area Foundation; formerly Alternative Energy Resources Organization, Executive Director

Larry S. Lev, Oregon State University, Extension Specialist and Associate Professor of Agricultural and Applied Economics

Harry MacCormack, Sunbow Farm, Owner-operator; Oregon Tilth, former Research and Education Director

Nancy Matheson, Alternative Energy Resources Organization (AERO), Agriculture Program Director

Dario Menanteau-Horta, University of Minnesota, Professor of Sociology; Center for Rural Social Development, Director

Helene Murray, University of Minnesota, Coordinator of Minnesota Institute for Sustainable Agriculture

Christopher J. Neher, Bioeconomics, Inc., Research Economist and Office Manager

David O'Donnell, Iowa State University, Research Assistant in Sociology

Susanne Retka-Schill, Northern Plains Sustainable Agriculture, Executive Secretary

Ron Rosmann, Practical Farmers of Iowa

Mary Lynn Roush, Oregon State University, Assistant Professor of Forest Science

Barbara R. Rusmore, Alternative Energy Resources Organization (AERO), Consultant and Director, Education for Democracy

James R. Sims, Montana State University, Professor of Cropping Systems

Melvin J. Stanford, Mankato State University and University of Minnesota, Visiting Professor of Agricultural Management

Karl N. Stauber, U.S. Department of Agriculture, Deputy Under Secretary for Policy and Planning, Small Community and Rural Development

David L. Watt, North Dakota State University, Associate Professor of Agricultural Economics

George A. Youngs, Jr., North Dakota State University, Associate Professor of Sociology

Barbara Weisman, formerly University of Minnesota, Research Assistant in Geography

Malvern Westcott, Montana State University, Western Agricultural Research Center, Superintendent and Professor of Soils

ABOUT THE NORTHWEST AREA FOUNDATION (NWAF)
AND THE CENTER FOR RURAL AFFAIRS (CRA)

The *Northwest Area Foundation* sponsored the research on which this book is based. The Foundation was established in 1934 by Louis W. Hill, son of James J. Hill, pioneer builder of the Great Northern Railway. Originally incorporated as the Lexington Foundation, it became known as the Louis W. and Maud Hill Family Foundation following the death of Mr. Hill. In 1975, the name was changed to the Northwest Area Foundation to reflect the Foundation's commitment to the region that provided its original resources and its growth beyond the scope of the traditional family foundation.

The Foundation takes its identity from this region, and agriculture emphatically is part of that identity. Six of the states are twice as dependent on agriculture as the national average, and 60 percent of the region's area is farmed. Farming and ranching still form the economic and cultural heart of many small-to-moderate-size communities in the region.

The Foundation's mission is to contribute to the region by promoting economic revitalization and improving the standard of living for its most vulnerable citizens. Currently, the focus of the Foundation's efforts is on alleviating rural and urban poverty and promoting sustainable development.

With current assets of approximately $300 million, the Foundation grants $12-14 million annually to nonprofit organizations in its region. It provided approximately $4.5 million over six years to fund projects in sustainable agriculture, including the research that has culminated in this book.

The *Center for Rural Affairs* was formed in 1973 by rural Nebraskans concerned about the decline of family farms and rural communities. CRA is a private, nonprofit corporation headquartered in the small farming community of Walthill, Nebraska. Its purpose is to provoke public thought about social, economic, and environmental issues affecting rural America, especially the Midwest and Plains regions.

The Center is committed to building sustainable rural communities, consistent with social and economic justice, stewardship of the environment, broad distribution of wealth and opportunity for all people to earn just incomes and own and control productive resources. CRA engages rural people in evaluating trends and policies in relation to these values and beliefs. The Center works through research, advocacy, education, organizing, leadership development, and demonstration of positive, holistic approaches to agriculture and rural community development.

PART

I

CRITICAL CHOICES FOR THE FUTURE

Why should you care about agriculture? Because agriculture is critical to our well-being, and it is in trouble. The legacy of conventional agriculture is good news—great plenty and advancing technology, but also bad news—environmental damage, concentration of production, and decline of family farming and rural communities.

A potentially beneficial alternative to conventional agriculture is emerging: sustainable agriculture. Part I provides a brief introduction to the issues that spawned the research described in this volume.

1

The Promise of Sustainable Agriculture

Karl N. Stauber
Chuck Hassebrook
Elizabeth Ann R. Bird
Gordon L. Bultena
Eric O. Hoiberg
Harry MacCormack
Dario Menanteau-Horta

Industrialized chemical-based agriculture has brought great bounty to America—and great problems of environment, economics, and social dislocation. An alternative farming strategy called sustainable agriculture promises a harmonious remedy to the problems—if it can be proved viable and become widely accepted.

WHY IS AGRICULTURE A CONCERN?

Agriculture is part of America's trusty infrastructure, so fundamental that it often is ignored. But here are important reasons why agriculture deserves our attention:

1. *Agriculture is the largest U.S. land user today.* Approximately 930 million acres, or over two-fifths of our land, is dedicated to food and fiber production (U.S. Department of Agriculture, Economic Research Service 1993, 5). Any enterprise that occupies this much real estate deserves scrutiny.

2. *America's agricultural bounty has hidden costs.* Our cheap food—costing consumers the smallest proportion of average income in any industrialized nation—actually is quite expensive if you include agricultural subsidies, the cost of long-term environmental degradation, and the cost of displaced farm families and declining rural communities.

3. *Multinational agribusiness exercises great control over farm production practices, commodities, and markets.* Americans need to recognize the extent to which agriculture is controlled by large corporations, and the implications—good and bad—of this control

of food pricing, quality, availability, environmental impact, and social impact.

4. *Agricultural chemicals and practices can have far-reaching effects.* They influence food quality, drinking-water quality, air quality, and the soil on which we live and grow food.

5. *Agricultural programs consume billions of tax dollars each year.* Much of this expenditure is toward dubious ends. A substantial share goes to help large farms increase in size at the expense of opportunities for small and beginning farmers. The rules by which subsidies are distributed encourage greater use of chemical pesticides and fertilizers.

6. *The massive displacement of rural Americans to cities affects us all.* Since 1910, rural Americans have been moving to cities for jobs. American agricultural policy stimulates larger farms, but their mechanization reduces employment, so this policy is forcing people to move to urban centers. Today, there are many urban homeless, while the homes left behind in small towns and rural areas are decaying.

7. *The U.S. system of agriculture is the basis of American civilization.* Farming displaced Native American civilizations, caused deforestation, helped establish democratic principles, helped overthrow European control, helped construct public transportation and finance, and helped settle every state. Those who care about where America is headed are wise to recall where we came from.

What is the role of farming and ranching in America's future? Are Americans locked into a farming system that will further consolidate control into the hands of an ever-shrinking group of profit-maximizing big farmers? Will only 4 percent of American farms produce 75 percent of all U.S. farm products, as predicted by the U.S. Congress Office of Technology Assessment (1986, 9)? Or is there an alternative view of future American farming?

THE LEGACY OF CONVENTIONAL AGRICULTURE

Our prevailing agricultural system, often called "industrial farming," has brought great bounty. It is characterized by rapid technological innovation and application, large-scale farming operations, highly specialized enterprises, large capital investment, high labor efficiency, and extensive dependency upon agribusiness. Agribusiness includes firms that collectively manufacture, process, and distribute farm products (Strange 1988).

The structure and practices of industrial farming align with stalwart American beliefs and values:

- Continued economic growth is necessary and desirable.

- Expanded productivity is essential to ensure abundant, cheap food.

- Larger farm units and improved labor efficiency are key to continued agricultural modernization and farm profitability.

- Technological innovation is an appropriate measure of agricultural progress.

- Profit and production maximization should be primary goals of farm operators.

In fact, the continued evolution of the industrial farming model is viewed by some as inevitable and essential to agriculture's future (Urban 1991).

But industrial farming's dark side is troubling:

- With the industrialization of agriculture have come increased environmental problems, including excessive topsoil erosion, water pollution, depletion of aquifers (underground rock strata that store water), and loss of wetland, prairie, woodland, and wildlife habitat.

- Industrialization has led to massive displacement of farm families. Since 1940, the number of farm dwellers in the states comprising the principal focus of this study—Iowa, Minnesota, Montana, and North Dakota—has declined precipitously, as shown in Figure 1-1.

- The precipitous decline in farm population comes largely from farm consolidation into ever-larger units. In three of our four study states, the number of farms now is fewer than half the count in 1940, and their size has grown concurrently—see Figure 1-2. Across all four states, average farm size (acreage) has approximately doubled during the past fifty years.

- Other worries include the recurring financial crises in agriculture, diminished autonomy and financial independence of farmers, high concentration of food and fiber production, widening financial disparity among farmers, declining opportunities for young people to enter agriculture, and the burgeoning power of corporate agribusiness.

It should be noted that many industrial agriculture proponents view

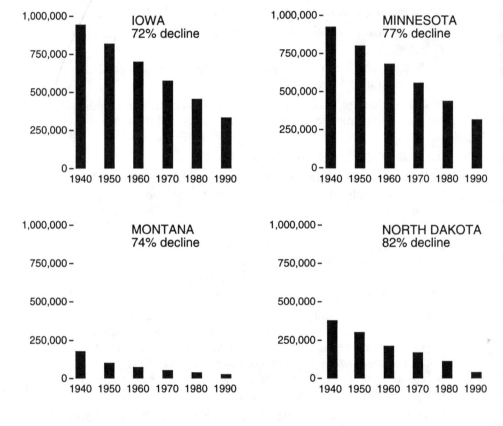

Figure 1-1. Changes in Farm Population, 1940-1990
(Source: Population censuses for the periods indicated).

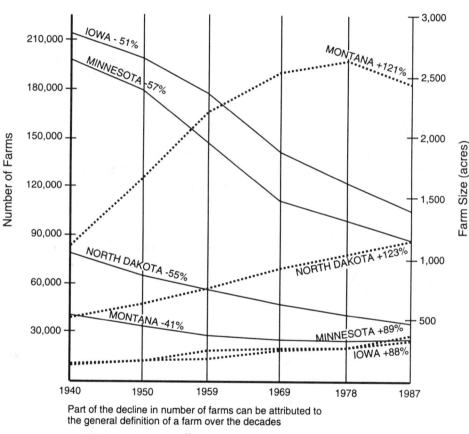

Figure 1-2. The trend: Fewer farms, larger farms. Percentages are change from 1940 to 1987. (Source: Population censuses for the periods indicated).

farm consolidation and the decline in farm population decline as positive trends, because they reflect technological innovation and increased labor efficiency. Farm depopulation also releases farm labor for reemployment—if jobs are available—in other sectors of the economy. In fact, increased labor efficiency and farm consolidation often are deemed essential if U.S. agriculture is to remain competitive in global markets.

Environmental Challenges

The environment rules farming: sunshine, moderate rain, and gentle slopes aid agriculture, whereas harsh winters, floods, drought, disease, and steep slopes hinder agriculture. However, farming technology has grown so powerful that modern farming techniques now threaten the environment with chemical contamination of drinking water (from pesticide, herbicide, and fertilizer), water depletion, excessive soil erosion, and degradation of wetlands and wildlife habitat.

Heavy use of artificial fertilizers and livestock confinement can contribute excessive nitrates to soil, groundwater, and streams. More than one-third of all counties nationwide—and a much higher proportion in agricultural regions—are vulnerable to groundwater nitrate pollution (Nielsen and Lee 1987). Excessive nitrates pose health hazards, especially to infants, to whom the effects can be fatal.

Agriculture not only pollutes water, but uses three times more than the commercial, industrial, and residential sectors combined. In some areas, this has led to groundwater depletion, conflict over water rights, and loss of fish species.

Soil erosion remains a vexing problem after a half century of federal soil conservation programs. As of the late 1980s, nearly a quarter of U.S. cropland was eroding fast enough to threaten long-term productivity. For example, Iowa has lost about half of its topsoil over the last 150 years.

Economic and Social Challenges

As farmland and farming opportunities have become concentrated into fewer and larger operations, many farm communities have stagnated or declined. Midwestern counties that depend on agriculture are losing population and typically suffer high poverty rates (Strange et al. 1990).

Declining farm communities mirror a decline in family farm opportunities and in the number of farming counties. Less than two percent of the U.S. population lives on farms today, compared to 25 percent in 1935 and nearly 50 percent over one hundred years ago. In 1950, over 60% of counties nationally were farming counties, compared to only

16% in 1986.

Without fundamental redirection, the population base of American agriculture will continue to decline. Nearly half of U.S. farmland is operated by farmers older than 55, and many will retire in the next decade (Smith 1988). There are fewer new farmers (Gale and Henderson 1991), and without a new farming generation, land worked by retiring farmers will be acquired by large, highly capitalized operations, with a permanent loss of small-farm opportunities.

As the number of farms declines, opportunities for farming self-employment are being replaced by contract-farming or wage-paying jobs on larger farms. During the 1980s, self-employed farmers represented fewer than half of the nation's farm workers, down from over 60 percent during the 1970s. These forces portend further social deterioration and increased social and economic inequality for the nation's agricultural communities (MacCannell 1983).

Advancing Technology, Decreasing Farm Share

Farm changes are driven largely by technology. Bigger and faster machines, greater reliance on chemical fertilizer and pesticides, and more energy input have increased capital requirements for farming, increased farm size, and often have replaced farmers' skilled labor and management. Importantly, the chemical and energy inputs are purchased from off the farm, becoming large expenses that divert substantial dollars away from the farm and create farmer dependency upon agribusinesses.

As technology has advanced, the farmers' share of the agricultural dollar has shrunk. From 1910 to 1990, the farm share of economic activity in agriculture declined from 41 percent to only 9 percent. During the same period, the share of the agricultural input sector and marketing sector grew correspondingly. Should this trend continue, the farm share of agriculture will be zero by the year 2020 (Smith 1992).

Public Policy on Agriculture

We emphasize that the federal government had no intention of increasing pollution, losing topsoil, and consolidating farms! However, federal policy bears part of the responsibility for the current state of agriculture in America. Federal policy was intended to raise farm income, stabilize farm prices, expand exports, ensure a safe and plentiful food supply, protect natural resources, and promote rural development.

The government has spent billions toward these often incompatible goals, mostly on the first three. But farm programs have encouraged larger farms, single-cropping (monoculture), larger equipment, more manufactured chemicals, and conversion of environmentally fragile land, such as wetlands, into crop production.

AN ALTERNATIVE: SUSTAINABLE AGRICULTURE

There is more than one way to farm the land. In America, we have been moving toward an industrial agriculture, but some farmers instead are moving in the direction of sustainable agriculture. What's the difference?

Industrial agriculture, in its most extreme form, has one focus: production. Its means are familiar: large acreages, powerful machines, potent chemicals to kill undesired pests, petrochemical fertilizers to promote plentiful growth, and billions of dollars in government subsidies to keep farms stable when shifting weather and markets threaten business.

Sustainable agriculture, on the other hand, works hard at stewardship. Generally, sustainable agriculture is diversified, flexible, environmentally sound family farming that replaces chemical-intensive practices with on-farm resources, renewable energy, conservation, and skillful management of natural processes. As shown in *Alternative Agriculture*, the National Research Council's ground-breaking book, some commercial farmers are succeeding with this new approach to farming (National Research Council 1989, 247-398).

UP FROM THE GRASSROOTS: SUSTAINABLE AGRICULTURE

To counteract the problems, sustainable agriculture emerged from the farm community as a grassroots initiative in the late 1970s. It began with farmers experimenting on their own farms to reduce the need for chemical pesticides and fertilizers through strategies that work in concert with nature, instead of in combat with nature. Examples of these strategies include:

- Greater crop diversity (planting a broader variety of crops)

- Crop rotation (instead of growing the same crop in the same field year after year, which depletes soil nutrients, this technique rotates crops through a multiyear pattern, such as corn-soybeans-corn-oats-hay, to replenish nutrients and control pests)

- Use of cover crops (for example, planting rye to hold the soil against erosion during a time when the field is not in other use)

- Use of soil-building crops (for example, planting legumes such as clover or alfalfa, which help restore nitrogen and organic matter to the soil)

- Intensively managed grazing systems (for example, systematically rotating livestock from paddock to paddock within a pasture to naturally recycle nutrients and minimize damage to the pastureland)

- Light cultivation for weed control (turning the soil to disturb weeds, instead of using herbicides)

- Integration of crops and livestock (raising both, and using each to support the other)

These innovations are being embraced by a small but important minority of the farm community, as well as by a growing network of grassroots organizations and agricultural scientists.

The pioneering farmers of sustainable agriculture not only use different practices but start from different assumptions. Typically, they hold a strong environmental ethic and serious concern about industrial agriculture's impact on the health of their families. They are an emerging group within the farming population that departs from industrial thinking about the goals, structure, production systems, and ideal characteristics of rural communities. (Please see Sidebar 1-1, "Two Farmers, Two World Views.")

SIDEBAR 1- 1

Two Farmers, Two World Views

Susanne Retka-Schill

Karl drives into the farmyard, enjoying the glorious spring morning. He commutes from town now, showing up to work at his farm each morning. It is easier for his wife and kids to participate in town activities now, since they built their new house there. The hired hand lives in the old farmhouse. The soil is drying quickly and he'll start planting in a couple of days, so his first job is to check over the new air seeder. The new seeder permits planting through more crop residue, and he can use his big four-wheel-drive tractor more efficiently.

"I've got good equipment," Karl thinks. "But I'm still experimenting with the crops. I don't like raising sunflowers—too many weeds. And I don't want to lose more farm program base acres—trying to fit those sunflowers in the farm program is too much of a hassle. But you know, another crop besides wheat and barley would be a good idea. I wonder what might fit in better."

Karl looks over the farm yard. The old barn his dad used for cows and

---

11

pigs was long ago converted into a shop and storage. Behind it, a few ducks paddle in what's left of the old slough. A few years ago, Karl drained it, and broke up the last of his dad's cattle pastures. "I'm glad I did that then," Karl thinks. "The farm program is getting tougher on drainage, and land is too expensive to waste on wetlands. Besides, as my equipment gets bigger, wetlands are just a nuisance."

Karl studies his field plans. The spring soil tests are back, and he can order fertilizer. Half of his ground got fertilized last fall, and he'll do the rest this spring. He used to apply the same amount to each field, but with low grain prices, he figures it's time to fine-tune applications. "I suppose I could cut back some—nitrogen levels might be too high—but I hate to lose yield." Karl takes pride in getting top yields for his area.

He looks over chemical literature. "That herbicide I've been using isn't working. This new one is supposed to be safer and give better weed control. I hope so—that weedy wheat field last year was a pain. Like Dad used to say, good farmers have clean fields—clean black soil with straight rows of grain. He'd be amazed at the trash we'll seed through with this new air-seeder."

Karl wonders about "no-till" farming. "If government rules gets tougher on crop residue and soil erosion, I may have to consider no-till. By tilling less, I could cover more ground, spreading my overhead over more acres to improve my net return. The only farmers getting anywhere these days are those that are getting bigger."

Karl looks across the farm yard. He has done well—good machinery, clean fields, a solid financial base. "Good management is the key to profitable farming these days," he thinks.

Across the county, John walks out the farmhouse door into the same beautiful spring morning. His farm is smaller than Karl's. "The fields are drying out, but I'll not seed yet—I'll wait until things are growing better— give that first flush of weeds a chance to sprout before I cultivate."

He starts the tractor to feed hay to the cattle. John no longer keeps a milk cow, like his dad, but he likes his beef cows. Its fun to watch the spring calves kick up their heels when they're turned out to pasture. "It will be nice to be done with chores. It should be dry enough to spread manure," he thinks. "I could get one pile spread today and work it in lightly tomorrow."

As he works, John weighs options . . . "My crop rotations are good. I do need another broadleaf plant, like sunflowers, to control weeds. I read about some new crops they are studying; maybe I'll try one.

"I'd sure like to rotate fields into pasture. They say grass is the best soil builder there is. I could add more cattle, but my buildings are just right for the ones I have. A better bet is sheep; they are good weed eaters. My wife wants to be more involved in the farm, so maybe sheep could replace her job in town."

John's thoughts turn to planting. Seed for the earliest fields was ready—wheat, barley, oats, and triticale, a rye-wheat hybrid. The flax and millet would go in last. "I need another crop for my rotation, but these specialty crops are hard to sell, and I can't risk losing more wheat acres in the farm program. I wish the farm program helped us diversify, instead of penalizing us. The rye I planted last fall looks good—maybe tomorrow I can broadcast sweet clover seed in it. Rye followed by clover, that ought to set back the thistles. I see why old-timers say good farmers have clean fields—it takes a good farmer to beat weeds ecologically!"

"More diversity would be good, maybe more livestock. Then I could farm smaller. Sometimes I think a good farmer is the one who can farm and live well on the fewest acres—that's efficient."

Over the years, John had developed a system that he felt was more sustainable—his crop yields were holding up, using natural fertility. Weeds were in check and seldom overtook a field. "We live comfortably here," he thinks. "Best of all, my kids can grow up like I did, with pets and animals and chores to do. Farm life is a good life."

* * * * *

Two good, thoughtful farmers. Two farming systems, Karl's moving toward the "industrial" model and John's striving for "sustainability." These hypothetical farmers display typical concerns of farmers who use each system. What might the future be like under each system? What are the policy choices through which we choose between them?

Sustainable farmers question these assumptions that undergird industrial agriculture: (a) nature is a competitor to be overcome; (b) progress requires unending evolution of larger farms and depopulation of farm communities; (c) progress is measured primarily by increased material consumption; (d) efficiency is measured by looking at the bottom line; and (e) science is an unbiased enterprise driven by natural forces to produce social good.

The philosophical underpinnings of sustainable agriculture include strong beliefs about:
- Integrating agriculture and nature.
- More localized and regionalized food systems to provide greater community autonomy and accommodate environmental constraints.
- Less centralized control of agriculture and farm resources.
- Greater self-sufficiency of farm operators.
- Independence of farm operators from the nonfarm sectors (chemical input and marketing).
- Greater cooperation among neighbors, and thus stronger farm communities.

The challenge of sustainable agriculture thus is social, economic, and environmental. In short, it is the challenge of sustainable development.

THE KEY POLICY QUESTIONS

Proponents of a more sustainable agriculture encounter vigorous skepticism—and often opposition—from those committed to the established farm structure and industrial practices. Among the key questions confronting policymakers and the farm community:

- Is sustainable agriculture a viable strategy for addressing the environmental, social, and economic challenges that confront farm communities?

- What actions should be taken by policymakers if sustainable agriculture is viable?

Policy decisions through which a sustainable agriculture strategy might be implemented are many:

- Should agricultural research and education programs start emphasizing the research knowledge and farmer decision-making skills required for sustainable farming?

- Should we redesign farm commodity programs to encourage sustainable agriculture and to remove the widely acknowledged barriers to it?

- Should agricultural policy actively encourage sustainable agriculture as a means of reducing environmental impact, or should it focus on improving industrial farming systems for that purpose?

- Should public policy enhance economic opportunities in family farming and farm communities, or should we as a nation abdicate family farm ideals?

    Public policy is a primary tool with which American society can select the future of agriculture, farm communities, and the rural environment. The research reported in this book is intended to help Americans use policy to make this choice.

IN THE FOLLOWING PAGES . . .

    Because context is essential to understanding our research findings, the next chapter is a brief agricultural geography of the region encompassing the five states involved in the Sustainable Agriculture Initiative and addressed in this book—Iowa, Minnesota, North Dakota, Montana, and Oregon.

In Part II, we report results from a coordinated research effort in four of these states—Iowa, Minnesota, Montana, and North Dakota. The research examined sustainable agriculture's implications for the structure of agriculture, farm families, and rural communities.

Part III describes lessons learned from on-farm research and agricultural ecosystems research. It reflects on the cross-institutional and interdisciplinary nature of our Initiative.

Part IV presents the research implications and the potential for public policy to simultaneously advance three things: rural community viability, family farm agriculture, and environmental quality.

Sidebars throughout the text focus on important project findings or help illuminate technical topics.

2

An Agricultural Geography of the Northwest Area

Gregory Chu
Barbara Weisman

The states involved in various aspects of the Initiative span the Northwest Area Foundation's funding region. The Northwest area is diverse in climate and soils, and consequently in its agriculture. A suite of maps depicts this diversity.

THE AGRICULTURE-DEPENDENT NORTHWEST AREA

The states involved in various aspects of the Initiative span the Northwest Area Foundation's funding region. The Northwest Area is the land once traveled by the Great Northern Railway, including Iowa, Minnesota, the Dakotas, Montana, Idaho, Washington, and Oregon. Of this land, 60 percent is in farms, about 17 percent more than for the country as a whole. More than half the counties in the Northwest Area depend on agriculture, compared to less than a quarter of all counties nationally.

WATER AND SOIL—THE NORTHWEST AREA'S VULNERABILITY

Agriculture varies dramatically across the Northwest Area due to uneven rainfall, temperature, growing season, and soils. In the west, dryland farming and irrigation prevail. Most irrigation water comes from rivers, streams, and reservoirs; the rest is pumped from underground aquifers. Crops in the more humid, eastern part of the Northwest Area are predominantly rain-fed, so just two weeks of below-average rainfall may cause drought. Long-term drought is a common threat further west, particularly in the dry Northern Plains.

At the other extreme, unusually abundant precipitation can cause flooding. During 1993, the Mississippi and Missouri Rivers and some tributaries inundated farmland throughout the Corn Belt and Northern Plains, causing the greatest crop failure in decades.

The detrimental impact of water and wind erosion on soil productivity long has been recognized. The Conservation Reserve Program (CRP) was begun in 1985 to reduce soil erosion by idling highly erodible cropland for at least ten years. The Northwest Area contains nearly 41 percent of the nation's CRP acreage, notably in the Iowa-Minnesota Corn Belt and wheat-growing Northern Plains and Palouse (eastern Washington).

AGRICULTURAL DIVERSITY

The region's agricultural diversity mirrors its geographic and climatic diversity. The valleys of western Oregon and Washington support more than a hundred high-value horticultural crops (fruits, nuts, and vegetables), whereas parts of Iowa and the Dakotas raise fewer than ten major crops or livestock.

Agriculture in the Northwest Area not only is a regional economic base, but also is a prominent player in the national economy, generating a major proportion of the country's leading grain and oil crop exports. Seven of the eight Northwest Area states are the sole or chief U.S. producers of specialty crops, such as potatoes, beans, peas, apples, berries, and hazelnuts.

The maps that follow portray the region's agricultural diversity, depicting commodities that are significant regionally, nationally, and internationally. Despite the variety of commodities shown, the true diversity of the region—actual and potential—is somewhat masked. Not visible, for example, is the market-gardening of fruits and vegetables throughout the region, nor other types of small-scale production.

These maps also cannot tell us which acres are being farmed sustainably. To convey the progress of sustainable agriculture in the Northwest Area would require collecting new types of data on agricultural systems and farming practices.

It is important to recognize that each map depicts an agricultural phenomenon that is the product of three things: natural factors (climate, soil, water, plant/animal adaptations), economic factors, and agricultural methods. The largely conventional agriculture portrayed is the realization of but one of many possible agricultural potentials. Over time, a more sustainable agriculture would add new products to these maps and shift the geography of production.

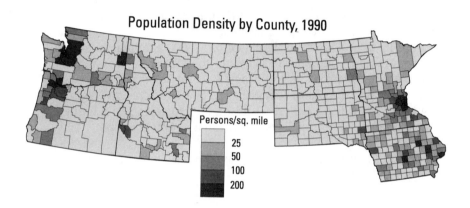

Figure 2-1. Northwest Area population. The unevenly distributed population is clustered at both ends of a sparsely settled interior. (Source: Bureau of the Census 1990c)

Figure 2-2. Farms and farmland in the Northwest Area.

(a) During 1982-1987, farm consolidation and other factors reduced the number of farms (mostly those of 50 to 250 acres) in Iowa and Minnesota by at least 5 percent, and at least 10 percent in about half of their counties. Conversely, during the same period several counties west of the Missouri River and into Montana increased farm count up to 9 percent.

(b) Farming is the region's economic base; more than half of the counties depend on agriculture, compared to less than a fourth nationally. In at least half the counties in Iowa and the Dakotas, about 90 percent of the land is farmed.

(c) The largest farms in the region are 2,000 acre-plus ranches in Oregon, Montana, and South Dakota. In the *Western Plains* and *Northern Plains,* farms range between 500 and 2,000 acres. *Corn Belt* livestock/grain farms range between 160 and 500 acres, averaging about 250 acres. The smallest farms, under 160 acres, grow horticultural specialties in river valleys of Oregon and Washington.

(Map sources: a: Bureau of the Census 1990b. b and c: Bureau of the Census 1990a)

Farms & Farmland

Number of Farms, 1987

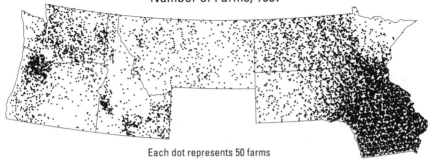

Each dot represents 50 farms

Percentage of Land in Farms by County, 1987

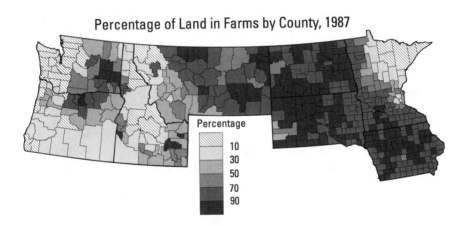

Percentage

10
30
50
70
90

Average Farm Size by County, 1987

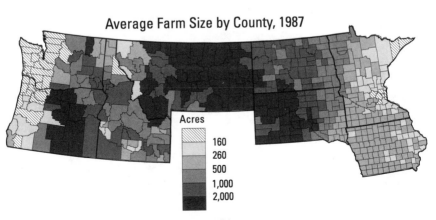

Acres

160
260
500
1,000
2,000

Figure 2-3. Precipitation, growing season, and soil fertility.

Spatial and temporal variations in these factors help explain the Northwest Area's current agriculture and suggest the region's potential for even greater diversity.

(a) Because all water for agriculture begins as rain and snow, precipitation profoundly influences farming in the region. East of the 20-inch precipitation boundary, farmers rely on snowmelt, followed by spring and summer rains, to keep the soil moist enough for crops. In arid/semiarid climates west of the 20-inch precipitation boundary, month-long dry spells warrant irrigation (the largest water consumer nationally). Paradoxically, flooding also is common in the Northwest Area.

(b) Crops need about three consecutive months above freezing to mature, so the growing season is the "freeze-free period" between spring's last killing frost and autumn's first. An adequate growing season throughout most of the Northwest area is modified regionally by climate and landscape.

(c) These extremely generalized mapping units show only *relative* fertility. Soil fertility is highly variable at the local level, and thus is not as uniform as this small-scale map implies. Some of Earth's most fertile farmland lies in the Upper Mississippi Valley and in the Palouse region of eastern Washington.

(Map sources: a: U.S. Council on Environmental Quality 1989. b: U.S. Geological Survey 1970. c: USDA 1981b and USDA Soil Conservation Service 1975)

Average Annual Precipitation

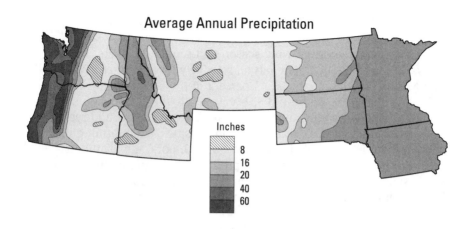

Inches

- 8
- 16
- 20
- 40
- 60

Duration of Growing Season

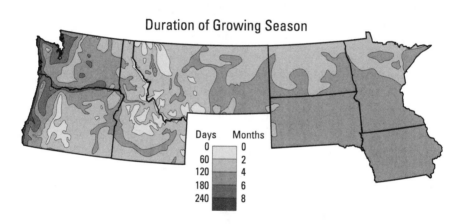

Days	Months
0	0
60	2
120	4
180	6
240	8

General Soil Fertility

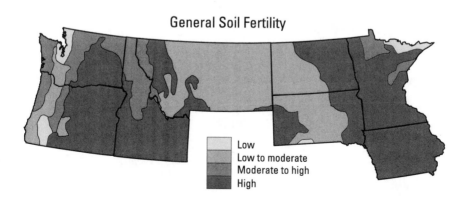

- Low
- Low to moderate
- Moderate to high
- High

Figure 2-4. Grain crops of the Northwest Area.

Tolerant of a short growing season and semiarid climate, wheat grows well in the moderately fertile, relatively level *Northern Plains.* The Northwest Area produced 43 percent of the U.S. wheat crop in 1991, with North Dakota the U.S. leader in all wheat production. That state also leads in barley, and South Dakota leads in oats and rye.

Corn Belt agriculture is combination livestock/grain farming that comprises an increasingly integral part of U.S. pork and beef industries. Iowa leads the nation in corn production. A moister climate and fertile soils are keys to Corn Belt agriculture. Corn production correlates geographically with soybean and hog production (see Figure 2-5).

(Map source: All map data from 1993 state agricultural statistics except for rye, from Bureau of the Census 1990b)

Grain Crops

Wheat

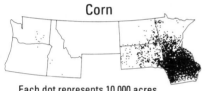

Each dot represents 10,000 acres
harvested for grain in 1992

Rye

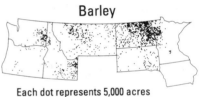

Each dot represents 2,000 acres
harvested for grain in 1992

Corn

Each dot represents 10,000 acres
harvested for grain in 1992

Barley

Each dot represents 5,000 acres
harvested for grain in 1987

Oats

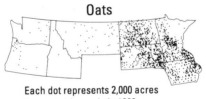

Each dot represents 2,000 acres
harvested for grain in 1992

Figure 2-5. Oilseed crops and major livestock inventories.

The *Northern Plains* region, especially North Dakota, leads the U.S. in sunflower and flaxseed. In soybeans, Iowa ranks second only to Illinois. Iowa, Minnesota, and South Dakota produced about a third of the 1991 U.S. soybean crop.

Large ranches dominate the rolling *Western Plains* (eastern Montana and western South Dakota). About half of Montana's farm-product income is from cattle and calves, but sheep are also important. In the *Northern Plains,* cattle graze about half of the land. In the *Corn Belt,* Iowa leads the nation in red meat production (beef, lamb and pork); Minnesota ranks third in hogs. Hog production and pork packing are heavily concentrated around Des Moines, and beef cows are concentrated in hilly, erosion-prone areas of Iowa that are less suited to corn and soybeans.

(Map source: Bureau of the Census 1990b)

Oilseed Crops

Soybeans

Each dot represents 10,000 acres
harvested for beans in 1987

Sunflower

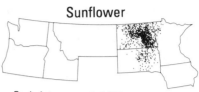

Each dot represents 2,000 acres
harvested for seed in 1987

Flaxseed

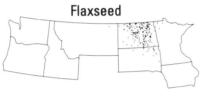

Each dot represents 2,000 acres
harvested for seed in 1987

Major Livestock Inventories

Hogs & Pigs, 1987

Each dot represents 5,000 animals

Sheep & Lambs, 1987

Each dot represents 5,000 animals

Milk Cows, 1987

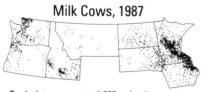

Each dot represents 1,000 animals

Beef Cows, 1987

Each dot represents 1,000 animals

Figure 2-6. Other important field crops.

Straddling the North Dakota-Minnesota border, the *Red River Valley* is noted for sugarbeets, dry beans, and potatoes. The Minnesota side ranks first nationally in sugarbeets; the North Dakota side ranks first nationally in dry beans (pinto and navy beans). In 1991, the Northwest Area grew just over one-third of the dry beans and nearly two-thirds of U.S. sugarbeets. Oregon's *Willamette Valley* is the country's number-one peppermint source.

(Map source: Bureau of the Census 1990b)

Other Important Field Crops

Potatoes

Each dot represents 1,000 acres
harvested in 1987

Sugarbeets

Each dot represents 1,000 acres
harvested in 1987

Dry Beans

Each dot represents 1,000 acres
harvested in 1987

Dry Peas

Each dot represents 1,000 acres
harvested in 1987

Lentils

Each dot represents 1,000 acres
harvested in 1987

Hops

Each dot represents 500 acres
harvested in 1987

Mint

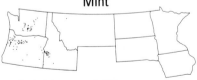

Each dot represents 500 acres
harvested in 1987

Figure 2-7. Major vegetable-growing areas.

The Northwest Area leads nationally in green peas, sweet corn, and onions. Minnesota leads the U.S. in green peas for processing, and is second, behind Wisconsin, in sweet corn for processing. These high-value vegetable crops generally are grown under irrigation by the companies that process and market them.

(Map sources: All data from Bureau of the Census 1990a, 1990b, and USDA Soil Conservation Service 1981, except carrots, from Bureau of the Census 1990b and USDA Soil Conservation Service 1981)

Major Vegetable-Growing Areas

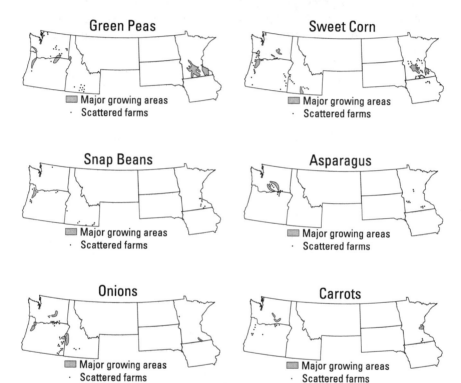

Green Peas
Major growing areas
· Scattered farms

Sweet Corn
Major growing areas
· Scattered farms

Snap Beans
Major growing areas
· Scattered farms

Asparagus
Major growing areas
· Scattered farms

Onions
Major growing areas
· Scattered farms

Carrots
Major growing areas
· Scattered farms

Figure 2-8. Major areas for tree and vine fruits, nuts, and berries.

Oregon's *Willamette Valley* is the number-one U.S. source of raspberries, blackberries, boysenberries, and loganberries, and virtually the only domestic source of hazelnuts.

(Map sources: Apples, pears, and grapes from Bureau of the Census 1990a, 1990b, and USDA Soil Conservation Service 1981; cherries, berries, and hazelnuts from Bureau of the Census 1990b and USDA Soil Conservation Service 1981)

Major Fruit-Growing Areas: Tree & Vine Fruits

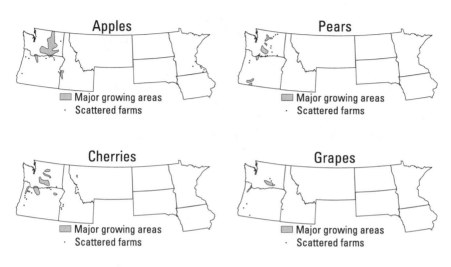

Apples
Major growing areas
· Scattered farms

Pears
Major growing areas
· Scattered farms

Cherries
Major growing areas
· Scattered farms

Grapes
Major growing areas
· Scattered farms

Major Fruit-Growing Areas: Nuts & Berries

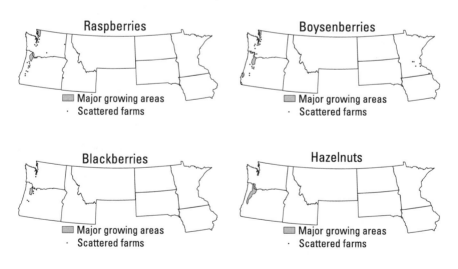

Raspberries
Major growing areas
· Scattered farms

Boysenberries
Major growing areas
· Scattered farms

Blackberries
Major growing areas
· Scattered farms

Hazelnuts
Major growing areas
· Scattered farms

Figure 2-9. Soil erosion by water.

This map shows water erosion, not only on cropland but on *all* nonfederal rural land. Estimates of soil loss by water erosion for a given area are based on rainfall, slope characteristics (length, shape, and steepness), the degree to which the land is protected by plant cover, and the way the land is farmed and managed, as well as the soil's inherent tendency to erode. Water erosion is severe in the *Corn Belt* and parts of the *Columbia Basin*. Wind erosion (not shown) is a serious problem in the dry, level *Northern Plains*. Whereas wind and water are physical phenomena that induce soil erosion, specific soil properties such as texture and water-holding capacity determine a soil's inherent vulnerability.

(Map sources: USDA Soil Conservation Service, National GIS Applications Lab 1993; USDA Soil Conservation Service National Resources Inventory 1987)

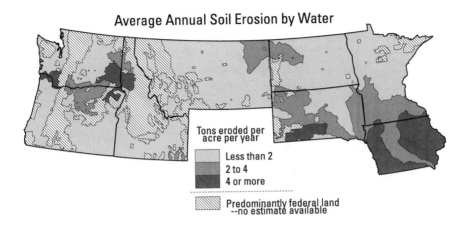

PART

II

SUSTAINABLE FARMING, SUSTAINABLE COMMUNITIES

Growing public disenchantment with adverse environmental, economic, and social impacts of industrial (conventional) agriculture has brought a recent appeal for alternative farming practices that are more sustainable. But there is vigorous debate about the relative merits of conventional and sustainable agriculture. Part II presents findings from a four-state study that systematically tested some of the claimed socioeconomic impacts (benefits and costs) of replacing conventional agricultural production systems with more sustainable ones. The research was performed as part of the Northwest Area Foundation's Sustainable Agriculture Initiative.

INTRODUCTION

Gordon L. Bultena
Eric O. Hoiberg

In the twilight of the twentieth century, public confidence in American agriculture is faltering. Accumulating evidence of adverse environmental, economic, and social impacts of industrial agriculture is provoking hard questions and rethinking of how best to produce our food and fiber. As a result, American agriculture is under mounting public pressure to adopt more sustainable production systems. But vigorous debate, both scientific and public, ensues about the relative merits of industrial farming and sustainable farming.

INDUSTRIAL FARMING VERSUS SUSTAINABLE FARMING

Proponents of industrial farming say it has proven highly productive and efficient, and its merits are long-established and speak for themselves. But they perceive many potential negatives of sustainable agriculture, claiming it will:

1. Reduce crop yields.
2. Diminish farm profits.
3. Raise food cost.
4. Reduce agricultural exports.
5. Require more tedious and costly farm labor.
6. Depress farm economies.
7. Undermine agribusinesses and diminish the economic viability of farm communities.

However, there has been little scientific appraisal of these claims. Recent study suggests that some of the negative predictions made for sustainable agriculture are inaccurate or overdrawn (National Research Council 1989).

Proponents of sustainable agriculture cite its potential positives, claiming it will:

1. Promote survival of family farms and slow (possibly reverse) the decline of the farm population.
2. Boost farm profitability and promote greater equity among farmers.
3. Help farmers become more self-reliant and less dependent upon agribusiness.
4. Provide fuller employment for farmers and farm families and use more farm labor to add value to agricultural products, retaining this value in rural economies.
5. Enhance job satisfaction, reduce farm-related stress, and improve

39

quality-of-life for farm families.

6. Strengthen local social institutions and community organizations.
7. Revitalize depressed rural economies.
8. Make U.S. agriculture less dependent on natural resources that are not renewable.
9. Safeguard environmental quality.

However, these claimed benefits of sustainable agriculture also remain largely undocumented.

A SOCIOECONOMIC ASSESSMENT OF SUSTAINABLE AGRICULTURE

To date, most comparisons of industrial and sustainable agriculture have focused on agronomic, economic, and environmental outcomes. Less attention has been paid to the socioeconomic consequences of a more sustainable agriculture. These include its effects on farm structure (size, ownership), farm enterprises (types of crops and livestock produced), farm operators and their families, rural communities, and rural culture. (Considerable speculation about socioeconomic impacts has been published; examples are Berry 1977; Bultena and Leistritz 1990; Kirschenmann 1992; Lasley, Hoiberg, and Bultena 1993.)

The Sustainable Agriculture Initiative's centerpiece was a four-state study (in Iowa, Minnesota, Montana, and North Dakota) that tested some of the claimed socioeconomic impacts of sustainable agriculture. Similar methods and survey questions were used in the four states to permit comparisons.

This four-state approach provides a more adequate test of social impacts than is possible from single-state research. Also, because these four states have such diverse ecological patterns and farming systems, we can test for differences in the impacts of sustainable agriculture among different locales.

This investigation was innovative in three important respects: (a) we developed operational measures for both sustainable and conventional practices; (b) we secured quantitative data on the sustainability of farms in diverse geographical regions that have widely varying crop and livestock enterprises; and (c) we objectively tested some purported socioeconomic impacts of sustainable agriculture.

We report the procedures and findings of this four-state study in Part II, following this outline:

Chapter 3—What Is Sustainable Agriculture? presents the definitions we developed to distinguish the two farm types (called conventional and sustainable) so that we could test for socioeconomic impacts; explains how the survey respondents were selected and data were collected; and describes the difficulties of measuring sustainability, especially in light of the marked diversity of cropping and livestock systems in the four states.

Chapter 4—Impact of Sustainable Farming on the Structure of American Agriculture describes how the characteristics of American agriculture might change if sustainable farming practices were widely adopted. Some characteristics of conventional and sustainable farms are compared, including size of operation, ownership, assets, sales, and types of cropping and livestock enterprises. Several personal and familial characteristics of the two farm types also are examined.

Chapter 5—Farm Labor and Management describes characteristics of labor and decision-making on the two types of farms; examines their respective farm-management requirements; and presents findings on labor source and amount, involvement of family members and others in decision-making, off-farm employment of the operators and their spouses, and managerial problems that arise in implementing more sustainable practices.

Chapter 6—Economic Position and Performance of Sustainable Farms compares productivity and financial performance of the two farm types (considering crop yield, farm profitability, and family income) and reports several indicators of economic performance for the two farm types (including gross farm sales, net farm income, and rates of return).

Chapter 7—Community Trade Patterns of Conventional and Sustainable Farmers compares the purchase locations of several production inputs and consumer goods by each farm type, and the extent to which their purchases are made in local communities or in more distant locales.

Chapter 8—Sustainable Agriculture: A Better Quality of Life? examines whether the two farm types display different levels of job satisfaction, farm-related stress, and farm family participation in community activities.

Chapter 9—Adoption of Sustainable Agriculture presents the prospects for widespread adoption of sustainable practices, including what motivates farmers to make the change, barriers they encounter, availability of information and advice needed to make a successful transition, and their satisfaction with choosing sustainable farming. The chapter also examines the contribution of sustainable farming organizations to stimulating, educating, and reinforcing farmers in their decisions to use more sustainable practices.

A BALANCED VIEW

As social scientists, we have tried to look objectively at the two farming systems, and to present a balanced view of each, both the positives and negatives. The impacts we studied were drawn from the literature on sustainable agriculture and writings of industrial farming proponents.

As attested by the diverse socioeconomic impacts studied here, the sustainable agriculture movement represents much more than mere promotion of alternative farming practices. It also seeks fundamental social and economic reform in rural society.

PROJECT PERSONNEL AND CONSULTANTS

Many persons were involved, in varying capacities, in the preparation, administration, analysis, and documentation of the surveys reported in Part II. Although common questions were prepared for use in the four states, it was the responsibility of those in each state to develop an appropriate sampling design and instruments for measuring farm sustainability and to collect and analyze the data. Data for the states then were shared and the authors listed for each chapter took responsibility for preparing the materials for publication. Thus, credit for the content of these chapters extends beyond the listed authors.

Persons participating in the design and conduct of the study in each state were:

• **Iowa State University**—Gordon Bultena and Eric Hoiberg (principal investigators), assisted by Derrick Exner, Susan Jarnagin, Nora Ladjahasan, and David O'Donnell; Ronald Rosmann and Richard Thompson (Practical Farmers of Iowa) served as consultants.

• **University of Minnesota**—Dario Menanteau-Horta (principal investigator), assisted by Virginia Juffer and Jodi Dansingburg; Ron Kroese (Land Stewardship Project) served as a consultant.

• **Montana State University**—Keith Jamtgaard (principal investigator), assisted by Al Kurki, Nancy Matheson, and Alan Okagaki (Alternative Energy Resources Organization).

• **North Dakota State University**—David Watt and Gary Goreham (principal investigators), assisted by Bruce Dahl, Randy Sell, Larry

Stearns, and George Youngs; Fred Kirschenmann and Sue Retka-Schill (Northern Plains Sustainable Agriculture Society) served as consultants.

Chuck Hassebrook and Elizabeth Bird (Center for Rural Affairs, Walthill, Nebraska) coordinated and facilitated the study. Cornelia Flora (Virginia Tech), Willie Lockeretz (Tufts University), Tom Dobbs (South Dakota State University), and Chuck Francis (University of Nebraska) served as consultants to the Foundation and as participants in the initial development of the study and interpretation of results. Toni Genalo, Iowa State University Statistical Laboratory, helped format the survey instruments and developed the common coding system used in the four states.

Most of the sidebars accompanying the chapters in Part II come from data that were not obtained in the surveys. Rather, they use findings from collateral projects in the Initiative that instruct interpretation of the survey results.

More detailed presentations of the survey data from each state are contained in the following publications:

Iowa   *Socioeconomic impacts of a more sustainable agriculture in Iowa.* Ames, Iowa: Department of Sociology, Iowa State University. Forthcoming.

Minnesota   Menanteau-Horta, Dario, Virginia M. Juffer, and Bruce Maxwell. 1993. *Patterns and trends of sustainable agriculture: A comparison of selected Minnesota farmers.* St. Paul: Center for Rural Social Development, University of Minnesota.

Montana   Jamtgaard, Keith. 1992a. *Results from the Montana Agricultural Assessment Project—Phase II interviews with sustainable and conventional producers.* Bozeman, Montana: Department of Sociology, Montana State University.

North Dakota Goreham, Gary A., George A. Youngs, Jr., and David L. Watt. 1992. *A comparison of sustainable and conventional farmers in North Dakota, final report, Phase II of the Northwest Area Foundation initiative on sustainable agriculture.* Fargo, North Dakota: Departments of Sociology/Anthropology and Agricultural Economics, North Dakota State University.

3

What Is Sustainable Agriculture?

John C. Gardner
Keith Jamtgaard
Frederick Kirschenmann

Sustainable agriculture is a goal rather than a specific set of practices, which makes it difficult to define. Its hallmarks are observable, but vary from state to state. We discerned its key principles, and from these developed a working definition for this study.

In Chapter 1, we presented a generalized definition of **sustainable agriculture:** *diversified, flexible, cost-effective, environmentally sound family farming that replaces chemical-intensive practices with on-farm resources, renewable energy, conservation, and skillful management of natural processes.* We offered this definition to give nonscientists a handle on the concept.

Many have sought to define sustainable agriculture, including academics, farmer organizations, advocates, and legislatures (for literature reviews, see Francis and Youngberg 1990 and Allen et al. 1991). Some definitions focus only on *environmental* criteria, although most include an *economic* criterion, such as profitability.

A consensus is growing that "sustainability" also has a *social* dimension. In its weakest version, the social dimension is merely "social acceptability." The congressional definition of sustainable agriculture in the 1990 Farm Bill added considerable momentum to a stronger assertion: to be considered "sustainable," farming systems *must enhance the quality of rural life* (including family farm opportunity) as well as environmental quality, human health, and farm profitability.

In identifying the critical elements to include in our operational definitions, we drew upon Lockeretz (1988) and Keeney (1989). Our definition deliberately excludes both economic and social elements, precisely because we wanted to examine the *linkages* between the use of sustainable farming systems, as we define them here, and measures of economic viability and quality of life.

Other definitions of sustainable agriculture that have yet to gain wide currency include a broader concept of equitable social relations. This concept includes food access, labor rights, and race and gender justice (Allen and Sachs 1993), and humane treatment of livestock (Gips 1984). This broader definition is gaining ground, however, as groups that focus on concerns of consumers, farmworkers, minority farmers, and animal welfare build a common campaign for agricultural policy reform with environmental and family farm organizations (National Sustainable Agriculture Coordinating Council 1994). We did not attempt a systematic examination of these issues in this study.

For this research, we developed an operational definition of sustainable agriculture. Before presenting this definition, we need to explain the principles that underlie it.

PRINCIPLES OF SUSTAINABLE AGRICULTURE

Classic agricultural research uses the *test plot,* an area of soil to which some treatment is applied, and from which results are measured. If repeated and varied many times, such experimental results can be used to formulate an agricultural principle.

Identifying the principles of sustainable farming, however, requires a more *ecological* approach: observing interactions on a whole-farm scale, examining the processes, and then formulating principles. The sustainable agriculture principles described in this chapter were drawn from the studies that comprised the Sustainable Agriculture Initiative. (The ecological method is described in Chapter 12.)

Most widely used sustainable practices address problems of conventional production. For example, in the Plains of North Dakota and Montana, conventional wheat farming requires numerous trips across fields with expensive equipment, and entails turning the soil, which contributes to wind erosion. These problems are addressed by the alternative practice of no-tillage wheat production, which means no turning of the soil, fewer trips across fields, and reduced equipment demand. Another sustainable practice, ridge-tillage, is a way to reduce water erosion and control weeds with less herbicide.

To the farmer, sustainable practices are ones that cut cost or improve overall results. In comparing various conventional practices and sustainable practices, the ones farmers find most successful have similar ecological impact. Ecologically, these practices offer strategies for reducing dependency on purchased inputs such as fertilizers and pesticides, and for gradually initiating new production systems as effective substitutes.

Each day, farmers make critical decisions on which crops and animals will be produced, how much tillage will be used, how pests (weeds and insects) will be managed, what machinery and supplies will be purchased, and how labor will be secured. Because the ways in which these problems are solved varies widely in our study states of Iowa, Minnesota, Montana, and North Dakota, it is difficult to distinguish any but the most obvious common practices that contribute to a more sustainable agriculture.

We group these practices under two broad principles: reducing dependency on synthetic, commercially produced fertilizers and pesticides, and developing more "positive" ecological practices, such as crop rotations and livestock integration. (Please refer to Sidebar 3-1, "What Do Sustainable Farms Look Like?")

SIDEBAR 3-1

What Do Sustainable Farms Look Like?

John C. Gardner

To most people, sustainable farms are not easily identified. But there are important clues: use of ecological niches, movement of fences and livestock, crop diversity, and generally simpler facilities.

Farming Natural Niches Instead of Square Fields

In ecological terms, a niche is a distinctive habitat created by the land's slope, how much solar energy it receives, its relation to surface water and groundwater, its exposure to wind, or its vegetation. To the farmer, a niche is an area where crops grow and yield differently.

In rolling, humid regions like Iowa, careful use of niches can flag a more sustainable farm. Instead of row crops stretching seamlessly, the crops share the landscape with carefully placed ribbons of cultivated pasture or alfalfa. Perennials normally are planted on steep slopes and along natural waterways to hold the soil year-round. Row crops are planted in between, but on the contour to retain both soil and water in the fields. In contrast to square fields running up over hillsides, the ecologically attuned farmer manages each niche for its best use. This causes less environmental impact and usually is more productive, even in the short term.

Each niche is created by several factors and requires management unique to its setting. For example, the Northern Great Plains has different soil types. Uniformly farming these diverse soils in traditional 160-acre quarter-sections is labor-efficient, but it can open the land to wind erosion and inefficient use of soil productivity.

An approach in Montana has been to farm soils, not fields. This is done by calibrating fertilizer application to individual soil types across large, varied fields (Carr et al. 1991). Most sustainable farms now align

---

field boundaries more closely to soil types than to surveyed section lines. The layout of these farms is visually evident, farming natural niches instead of square fields.

In other areas, farmers create their own niches with selective plantings. In Oregon's Willamette River Valley, which has naturally fertile soils, high-value crops typically are grown within fields of uniform soil and uniform slope. Farmers create niches through the succession of diverse crops, grown intensively and close together. These ecologically managed fields are distinctive for their diversity and the visible care taken in crop arrangement.

On the sustainable farm, the landscape not only features the most adapted plant in each niche; the niches interrelate and are mutually supportive.

Movement of Fences and Livestock

Another visual clue to sustainable farming is the division of pastures with mobile fencing and temporary electric fencing, instead of traditional permanent fencing. In the Corn Belt, for example, pastureland increasingly is fenced for rotational grazing. The animals are rotated often to utilize their considerable impact on the land—grazing, manure deposition, and hoof action—in a cyclical, positive way.

Letting the land "rest" from livestock impact allows forage to recover and to hold the soil against erosion. It also allows time for weathering to incorporate manure nutrients into the soil. Yet another benefit is that it can break the life cycle of some pests. In effect, livestock rotation allows the animals to manage the plants. Farmers using rotational grazing are enjoying improved productivity, both in their animals and their pastures (see Sidebar 4-2, "Controlled Grazing").

Such rotation of livestock requires a quick, economical confinement method: movable fence. Such fencing is an important clue to a sustainable farm.

Crop Diversity

In the Northern Plains, crop diversity is essential to farmers who seek sustainability. Alternating with traditional spring cereal grains, you can see buckwheat, millet, winter rye, and a green-fallow legume. The reason, however, is not attractive markets for these crops. These crops are used because they are planted and grown during different periods of the year than the traditional cereals.

Such asynchrony among crops in the rotation is necessary to manage weeds and control diseases. It also spreads soil moisture consumption through more of the year, important because of water scarcity in this dry region.

In the Corn Belt, crop diversity is less critical because the climate is wetter and there are fewer economically competitive crops. Thus, crop diversity is a less important clue to identifying the sustainable farm in the Corn Belt.

Simpler Facilities

A more subtle difference is a tendency for sustainable farms to invest less in energy-intensive and capital-intensive buildings and enterprises. On a sustainable farm, sows are more likely to reside in individual housing on-pasture than in a slotted-floor, electrically ventilated confinement building. Sustainably raised beef cattle typically eat more forage than grain until their "finishing rations" fatten them for market. This means less investment in finishing facilities.

Personal housing, tractors, pickups, and combines vary as much on a sustainable farm as on a conventional farm. All farming displays normal human variety: some farmers like it new, but others old; some like it big, some small; some are neat, others are not.

Management

The most significant aspect of a more ecologically sustainable farm is the least observable: the difference in its management. Management is the fundamental key to identifying a sustainable farm, but it can be observed only indirectly, through a combination of niche use, repositioned fences, crop diversity, and simpler facilities. A farm managed with the goal of sustainability can be recognized if you know what to look for.

Reducing Dependence on Industrial Fertilizers and Chemicals

Conventional U.S. agriculture is very sophisticated at supplying soil nutrients and controlling pests. Fertility management commonly is guided by soil testing and purchase of nutrients (chemical fertilizer) to meet needs of specific fields. Weeds and insects are controlled through broad-spectrum pesticide applications, followed by field scouting and more targeted applications where needed.

Alternatives to these common chemical practices were found in each state. Although they vary by locality, they are similar in principle. The four basic alternatives for reducing dependency on chemical inputs are (1) use of manures for soil building, (2) tighter loops of nutrient cycling, (3) sharper scrutiny of chemical input needs, and (4) substitution of mechanical weed control for chemical weed control. We will briefly examine each.

Use of Green or Livestock Manures for Soil-Building. Crops deplete the soil of nutrients, which must be replaced. An alternative way to do this is with manure. There are two types: livestock manure and green manure crops.

Livestock manure requires a low-cost source, such as on-farm cattle or swine. *Green manure* is a crop planted specifically to improve soil fertility, and not for harvest. Common green manure plantings include sweet clover, lentils, and alfalfa. These plants are *legumes,* which add nitrogen to the soil. The plants enrich the soil as they grow, and continue to do so when plowed under.

Sustainable farms generally rebuild soil by including green-manure legumes in their crop rotation. In Montana and North Dakota, green-manured sweet clover, black lentils, or other legumes are grown every fourth or fifth year in the rotation (Figure 3-1). These not only provide nitrogen and other nutrients, but also conserve moisture for the ensuing crops when properly managed.

In the Plains, farmers long have practiced *fallow,* a year when a field is left unplanted and the soil is tilled frequently to conserve moisture and control weeds. The bare, dark soil gives this period its name: *black fallow.* On some sustainable farms, black fallow now is replaced with *green fallow,* during which a green manure crop is planted.

Black fallow and green fallow are similar in cost, and both conserve water if properly managed. But green fallow also provides plants to hold the soil and reduce erosion, and adds nutrients both while the plants are living and when they are plowed under. Another alternative that has become common in recent years is *brown fallow,* the use of herbicides to create a residue of dead plants to retain the soil.

In more humid regions, rotations commonly include legumes, but rely more on cash-producing crops (alfalfa, soybeans) and use of livestock manure than on green manure, which is not a cash-producing crop. In dry regions, when livestock manure is used, a little goes a long way. It usually is applied once every other rotation cycle. The greater rainfall permits more intensive rotations (no fallow period) but livestock manure usually is required every few years. Livestock in this region are either concentrated onto small acreages to make manure collection efficient, or are spread over intensively managed pastures to reduce the cost of manure storage and distribution.

Tighter Loops of Nutrient Cycling. Conventional practices emphasize maximum harvest from crops. This replenishes less organic matter than did the original grasslands. This is an "open system" that requires external chemical input to restore. It has led to many existing assumptions about soil fertility and management, which are true in the context of conventional production.

However, these assumptions do not always apply to natural ecosystems. Nor do they always apply to some of the tighter nutrient-cycling loops that occur on many of the sustainable farms studied in these states.

In the Prairie, perennial plants develop tremendous root systems. Thus, Prairie ecosystems store the majority of nutrients, energy, and biomass in their roots, within the soil itself. This offers safety and stability for surviving drought, fire, harsh winters, and grazing animals.

Similarly, sustainable farms purposely invest in the soil to stabilize and conserve nutrients.

Sustainable farmers use several techniques to develop a soil capable of supplying crop nutrients and retaining a sizable nutrient store. They regularly incorporate green or livestock manure; they produce high-residue crops, like tall wheat, which provides more straw to return to the soil; and they include in their rotations perennial forage grasses and legumes, which form extensive root systems.

Sharper Scrutiny of Input Needs. Many farmers who once used fertilizer and pesticide generously as preventives are reducing these purchased inputs. As an example, a late-spring soil test and a cornstalk nitrate test developed in Iowa help avoid excessive nitrogen fertilization. Farmers using these tests can maintain corn yields with less nitrogen than conventional soil-testing procedures recommend.

Substituting Mechanical Weed Control for Chemical Weed Control. Farmers in wheat- and corn-producing regions are using alternative tillage to reduce or even eliminate herbicide use. There has been a shift to additional tilling after planting, done very early when crops are freshly emerged, using special implements that disturb weeds more than crops, giving the crops a competitive advantage. In North Dakota and Iowa, this post-planting tillage now often requires more management than traditional pre-planting tillage.

Despite this extra work, sustainable farmers generally make fewer overall tillage passes across their fields than their conventional neighbors. The reasons are that less primary tillage is done, and sustainable farms use more biennial (two-year) and short-term perennial crops in rotation, which require less tillage.

Developing More "Positive" Ecological Practices

Increased specialization has been the trend over the past few decades, but it unwisely creates ecological imbalance and instability. Countering this trend is crop diversification, a common feature of sustainable farms. Well-planned and carefully implemented diversification provides stability, both ecologically and economically.

Regardless of location, sustainable cropping relies on substituting low-impact, long-term, generally stabilizing ecological processes for high-impact, short-term inputs. These are called *positive* practices, and they include crop rotation, integrating crops with livestock, and landscape management. Each is described below.

More Ecological Crop Rotation. Crop rotation to improve the soil is not new, but traditional rotations often fail to break pest cycles or to

51

provide as much soil rebuilding as they could. An improved sequence of complementary crops allows for mutually supportive relationships to develop in the soil. The key to effective crop rotation is *complementary plants.*

Rotation strategies used in the semiarid Plains include alternating deep-rooted crops with shallow-rooted crops, early-planted crops with late-planted crops, and warm-season crops with cool-season crops (Figure 3-1). In Iowa and Minnesota, different rotation strategies are being tried, with various combinations of corn, soybeans, alfalfa, oats, peas, and strip intercropping.

Conventional crop rotation—similar plants, similar pests and weeds, no soil enrichment:

Year 1	Year 2	Year 3	Year 4	Year 5
wheat	barley	wheat	black fallow	(repeat
(cash crop)	(cash crop)	(cash crop)	(no crop)	the cycle)

Sustainable crop rotation—dissimilar plants, breaks pest and weed cycles, enriches soil:

Year 1	Year 2	Year 3	Year 4	Year 5
spring wheat	winter rye	sunflower	green fallow	(repeat
(cash crop,	(cash crop,	(cash crop,	with sweet	the cycle)
shallow-rooted,	about 1½ feet	2-3 feet	clover	
cool season,	deeper-rooted,	deeper-rooted,	(no cash crop;	
early-planted)	cool season,	grows most in	deep-rooted,	
	fall-planted)	mid-late summer)	biennial)	

Figure 3-1. Examples of conventional and sustainable crop rotation in the Plains.

The alternating of winter-planted and spring-planted crops, or early-planted crops with late-planted crops, in combination with greater seeding rates, is the primary mechanism for weed control. These crop rotations create a "perpetually interrupted plant succession," which keeps weeds at a competitive disadvantage to the desired crop.

In all, these expanded rotations require a greater number of crops, or greater diversity. Nearly half of all farms using such rotations grow crops that are considered to be alternative, or specialty, crops for their region.

Integrating Crops with Livestock. Not all sustainable farms have livestock, but it is a common diversification strategy. Livestock production practices vary widely. Greater rainfall areas have *intensive* livestock enterprises (greater concentration of animals in smaller areas), whereas drier regions have *extensive* enterprises (animals spread over much greater areas).

In Montana, for example, cropping enterprises often are coupled to ranching enterprises, with grazing on poorer land. Also, grazing is integrated into the cropping operation on some farms that include pasture in their crop rotation.

By contrast, Iowa farms may use partial confinement of hogs or rotational cattle grazing. These practices support animal enterprises without sacrificing farmers' ability to plant annual field crops. Adding livestock can expand crop rotations by adding feed crops for the animals: grains, legumes, grass, and other hays.

Further, hay and silage crops aid in weed control because harvesting almost completely removes these plants, thus reducing weed seed in the field. And if these crops are damaged by drought or hail, they can be salvaged by feeding them to livestock. This is an indirect means of economically marketing crops that otherwise might be lost. (Silage is animal feed that consists of entire plants preserved through anaerobic fermentation or pickling in a silo. To make silage, farmers chop entire plants—for example, whole corn plants, not just the ears. Although this provides maximum feed from a field, it also returns little organic matter to the soil.)

Landscape Management. Greater landscape diversity is common on sustainable farms. Instead of vast fields divided by property boundaries or standard 160-acre quarter-section tracts, sustainable farmers plan their fields according to soil type, terrain, wind exposure, drainage, and so on.

Strip cropping is found on some sustainable farms. These strips, generally 200 to 300 feet wide, are perpendicular to the prevailing wind or slope to reduce soil erosion. Strip cropping adds biological diversity and often improves crop yield, while still using traditional equipment and field operations. In the Plains, strip cropping on sustainable farms now is the rule.

In the Corn Belt (Ohio, Indiana, Illinois, Iowa, southern Minnesota, and northern Missouri), strip cropping is less common, and uses much narrower strips. Alternating narrow strips of tall corn and short soybeans not only form windbreaks, but also create alternating microclimates in the rows, which help control plant diseases and insect pests.

Other landscape modifications are becoming more common, with the goal of improving sustainability. In Iowa, sustainable farms are using more trees for windbreaks and contoured grass waterways for erosion control. Previous techniques included building terraces or erosion barriers. Now, on sustainable farms, we note perennial trees and grass, cover crops, crop rotations, and greater use of crop residue to achieve erosion control.

Two other management practices that seem to play an important role in influencing the landscape are reduced tillage and increasing crop biodiversity. Farms that practice conservation tillage reduce soil erosion by maintaining year-round surface cover with crop residues. ("Conservation-till" is a practice of conserving soil through the reduction or elimination of tillage. It generally includes a shift from deep plowing to shallower surface tillage, leaving more soil-holding residue undisturbed on the surface.) Most sustainable farms include a greater diversity of crop plants grown in rotation. The combination of year-round surface cover and greater plant diversity generally creates a more complex and stable biotic community.

Although little currently is known about the impact of such purposefully managed landscapes, we can assume that diverse mammalian and avian wildlife, insect predators and prey, and other organisms will promote a more stable, more resilient ecology to enhance long-term crop productivity.

Reducing Industrial Inputs and Implementing "Positive" Ecological Practices

While some farmers have reduced their purchases of chemical fertilizers, their reasons for doing so are varied. Some reduce purchases primarily from cost concerns, and do not attempt to replace these input purchases with "positive practices." We have observed that farming sustainably requires a high level of commitment to these positive practices, in addition to reducing input purchases. Such a commitment reflects the farmer's belief that farming sustainably is a worthy goal in itself.

The positive practices described here take longer to implement than conventional practices. Adoption of these new practices—more crop rotation, new tillage practices, adding livestock—often requires strong motivation and patience to carry out successfully. Most farmers who have adopted these practices did so slowly, deliberately, and with help from more experienced farmers. (For a look at "How Some Farmers Perceive the Impact of Sustainable Agriculture," please see Sidebar 3-2; also Chapter 9.)

SIDEBAR 3-2

How Some Farmers Perceive the Impact of Sustainable Agriculture

Eric O. Hoiberg

Gordon L. Bultena

Jodi Dansingburg

When Minnesota farmers were asked for their definitions of sustainable agriculture, conventional farmers generally defined it economically, emphasizing fewer purchased inputs, especially chemical pesticides and fertilizers. Sustainable farmers often defined it in environmental and community terms, as well as economic. One invoked the view of Wendell Berry, often considered a philosophical father of the sustainable agriculture movement, that sustainable agriculture is "a farming system that neither depletes the people or the land."

Some farmers (both sustainable and conventional) offered no definition. Either they were unfamiliar with the term, or did not understand the style of farming it implied. This suggests that sustainable agriculture has yet to become well understood by all farmers, despite increasing coverage in the farm press.

Intense debate swirls about sustainable agriculture's purported social, economic, and environmental impacts. Unfortunately, this debate seldom is informed by scientific evidence. So, we asked a cross-section of Iowa farm operators (55 sustainable and 52 conventional) this question: "If you were to adopt sustainable practices on your farm, what outcome would you expect?" They reacted to thirteen specific factors, spanning production, management, environmental, and personal/familial considerations:

Complexity, Labor, and Erosion. Conventionals and sustainables closely agree on some impacts, generally believing that sustainable practices mean increased management complexity, more labor, and less soil erosion and soil compaction. These polar groups also concur that farming sustainably is unlikely to affect their standing in their local communities.

Yield, Profit, and Risk. Conventionals and sustainables disagree on many other impacts. Conventionals generally feel that crop yields will decline; sustainables believe yields will be unaffected. Similarly, conventionals most often foresee declining profits; sustainables typically expect no change or even increased profitability. Conventionals expect greater financial risk; sustainables see this as unchanged or even diminished.

Family Health. Respondents disagreed on the likely impact of a more sustainable agriculture on family health. Virtually no one foresees increased family health problems from sustainable practices. Conventionals, more often than sustainables, expect little or no change in family health. However, sustainables more often foresee reduced health problems.

Career Satisfaction. How do respondents feel that sustainable farming will affect their job satisfaction? Not surprisingly, a pervasive view among sustainables is that their job satisfaction increased with adoption of the new practices. Only a small number of conventionals share this sentiment; most feel that their present level of job satisfaction, which generally is high, would be little affected by adopting the new farming practices.

These farmers' answers are instructive because their expectations clearly affect their decisions to adopt—or not to adopt—sustainable practices.

FOR FURTHER INFORMATION
Bultena, Gordon L., and Eric O. Hoiberg. 1992. Farmers' perceptions of the personal benefits and costs of adopting sustainable agricultural practices. Impact Assessment Bulletin 10(2): 43-57.

As described in the next section, the principles and commitment to them formed the foundation for the working definition of sustainable agriculture used in our surveys. Because of the marked variation in climate, soils, and agricultural practices across the four-state study region, operational definitions were adapted to each state's situation.

STUDY METHODS

(NOTE: This section briefly overviews the methodology used in the four-state study. Please see Appendix A for a more detailed discussion.)

Prior to the Sustainable Agriculture Initiative, most information about sustainable farmers had come from case studies of operators who are notably successful with sustainable practices, or from studies of sustainable farming organization members. However, the extent to which these persons are representative of the larger population of sustainable farmers is unclear.

Also, some researchers have defined sustainability too narrowly—for example, simply equating it with reduced input use. This ignores the important aspect of adopting more "positive" practices, such as replacing inorganic fertilizer with animal and/or green manure.

Consequently, our study differed from much previous research in two important ways: (1) its use of representative samples of farm operators in four states (Iowa, Minnesota, Montana, and North Dakota), and (2) its multidimensional definition of sustainability (rather than defining it by a single criterion).

The study was conducted in two phases. In Phase I, we selected representative farm operators for study. In Phase II, subgroups of conventional and sustainable farmers were selected from those chosen for Phase I.

Phase I

In Phase I, we contacted 3,567 farm operators, randomly selected from the general farm population in the four states. Of these, 2,450 (69 percent) provided information about their farming practices by telephone interview, mail questionnaire, or both.

Nearly a thousand individuals in the four states were identified by sustainable farming organizations as either practicing or interested in sustainable agriculture. Of them, 796 (81 percent) provided information. This *supplemental* sample helped insure that enough sustainable farmers would be available for analysis.

We were aware that those who belong to sustainable farming organizations may show differing patterns. [Please see Sidebar 6-2 ("In-Field Studies of Sustainable Farm Productivity") and Sidebar 9-3 ("What Role Do Sustainable/Organic Farming Organizations Play?").] However, the organization members were not defined as farming sustainably unless they met the same criteria as persons from the farm population samples.

In Appendix A, Tables A-1 and A-2 describe the study designs and Phase I samples.

A Working Definition of Sustainability. An important part of the Phase I data collection was measuring the sustainability of each respondent's farming operation. Because of the diversity of crops and farming practices across the study area, the specific nature of sustainability varied from state to state. However, we used three general principles to form a working definition of sustainability in all four states:

1. *A sustainable farmer substitutes farm-generated or locally available production inputs to replace inputs produced outside the area (such as commercial fertilizers and chemical weed controls).*

2. *A sustainable farmer has adopted "positive" practices that diversify farm operations and provide alternatives to synthetic fertilizer and chemical use.* Positive practices include substitution of animal and green manures for synthetic fertilizers, use of tillage and crop rotation, and biological control of pests rather than chemical treatments.

3. *A sustainable farmer is committed to using locally produced inputs, and to enterprise diversity.* Rather than perceiving sustainable practices as alternatives to use during difficult times, sustainable farmers are committed to changing their whole approach to agriculture, consistent with their new values.

Table 3-1. Operational measures of agricultural sustainability

Indicator	Iowa[1]	Minnesota	Montana[2]	North Dakota
Locally Produced Inputs				
Nutrient Practices	Pounds of crop-available nitrogen applied per acre (includes green and animal manures); percentage of all nitrogen fertilizer purchased	Most nitrogen not provided by commercial fertilizers; most acres rotated with legumes or green manure crops; dependence on organic/natural or synthetic/chemical fertilizers	Green manure crops; pounds/acre of chemical fertilizer applied to principal cash crop; microbial soil amendments; deep soil sampling; reduced primary tillage	Commercial fertilizers; animal manures; percentage of cropland used for green manures
Weed and insect control	Per-acre expenditures on herbicides, insecticides	Control weeds by tillage and other cultural practices with herbicides, or both methods equally	Nonchemical weed control; trend in pesticide use (increase, decrease, same)	Percentage of cropland on which herbicide is used
Energy use	Energy inputs (kilocalories per acre); cost of drying corn	(Not used)	(Not used)	(Not used)
Positive Practices				
Crop diversity and rotations	Percentage of cropland in crops other than corn or soybeans; number of crops planted in field during a five-year rotation	Number of crop rotations used	Diversify crops; use long-term crop rotations; number of years in crop rotation; number of crops reported	Number of different crops grown on over 5% of cropland
Livestock	Number of different livestock enterprises	(Not used)	Number of different animal species reported, rotate manure applications apply manure to farm land, integrate crop/animals, use hormones, use probiotics, intensive pasture rotation	(Not used)
Commitment				
Self Identification	With minor differences among states, respondents indicated agreement with one of three statements: one indicating reliance on purchased fertilizers and pest controls, one indicating attempt to decrease purchase of fertilizers and pesticides, and one indicating reliance on farm-based fertilizers and pest-control techniques			
Other assessments	Four evaluations of sustainable practices: three or more crops in rotations; reliance on pesticides; dependence on purchased inputs; nonchemical weed control	(Not used[3])	Personal involvement in marketing	One item measured agreement with use of chemicals and scientific advances; another measured agreement with reliance on external inputs and energy

[1] Iowa researchers developed and separately examined an index that measured percent changes in agricultural practices, although these data are not presented here.

[2] Listed here are all the items used among three different production systems. Not all of the items were used for each of the systems.

[3] Minnesota researchers developed a separate index to measure the attitudes of respondents regarding the perceived dangers of existing farming practices, and the desirability of alternatives.

Each of these principles was captured with multiple measures. The specific criteria used to measure sustainability in each state are shown in Table 3-1.

SIDEBAR 3-3

Which Practices Define Sustainability?

Sheila M. Cordray

Larry S. Lev

Richard P. Dick

Helene Murray

Horticultural farmers raise fruit and vegetables—for example, potatoes, broccoli, sweet corn, raspberries, and strawberries. In western Oregon and Washington, we surveyed some horticultural farmers to relate their range of production practices to sustainability. The most interesting finding was the varied dimensions of what we call "sustainable."

• When we referenced sustainability to reduced use of agricultural chemicals, those who reported declining or no chemical use tended to have smaller farms, less machinery investment, and lower gross and net incomes.

• When we referenced sustainability to reliance upon on-farm resources rather than purchased off-farm inputs, farmers using more on-farm resources had larger farms, more machinery investment, and larger gross and net incomes.

Clearly, these results conflict, demonstrating the difficulty of defining sustainability based on farmers' practices. Further, farmers classified as sustainable by our measures differed little from conventionals in community economic impact, organizational involvement, or attitude about farming.

The study findings suggest that structural factors such as size of farm and the principal occupation of the farmer can have an important influence on the production practices used. Policies to encourage farmers toward more sustainable practice should take into account the structural factors that influence this adoption.

FURTHER INFORMATION

Cordray, Sheila M., Larry S. Lev, Richard P. Dick, and Helene Murray. 1993. Sustainability of Pacific Northwest horticultural producers. Journal of Production Agriculture 6(1): 121-25.

Distinguishing "Conventional" and "Sustainable" Farmers. We scored each subdimension of sustainability in Table 3-1 and derived a cumulative score (a *sustainability index*) for each respondent. These cumulative scores reflect the relative sustainability of each operator.

Their sustainability index scores fell along a "sustainability continuum" from high to low:

- *Those having high cumulative index scores were labeled "sustainable" farmers for purposes of this study.* They typically apply comparatively low amounts of synthetic fertilizer and chemicals, have diversified operations with additional crops and livestock, are using "positive" practices, and perceive their sustainable practices as necessary and desirable.

- *Those having low scores were labeled "conventional" farmers for purposes of this study.* Conventional farmers simply are the polar opposite of sustainable farmers. Compared to farmers in general (and especially to sustainable farmers), "conventional" farmers use large amounts of crop nutrients and chemicals, have highly specialized cropping systems, and are skeptical (if not hostile) toward sustainable farming practices. Thus, "conventional" does not imply a residual category of farmers that remained after we selected sustainable producers, nor does "conventional" imply a cross-section of all farmers.

Issues in Identifying Sustainable Farmers

Sustainable Versus Organic. As defined in this study, *sustainable agriculture* is not interchangeable with *organic farming*, although the two terms sometimes are confused. Organic farming carries a strong commitment to avoid *all* chemical pest controls and inorganic fertilizers, whereas many sustainable farmers still use these inputs, but in reduced amounts. (Please see Sidebar 3-4, "Sustainable Farming Versus Organic Farming.") Organic farmers are represented in our sample, varying in number by state.

SIDEBAR 3-4

Sustainable Farming Versus Organic Farming

Harry MacCormack

Unlike "sustainable" farming practices, organic farming practices are well-defined—in fact, organic farming practices are unique, for they are the only ones codified as law. A complete set of certification procedures governs organic farming, from the soil to the dining table.

Thus, organic farming is clearly mapped for farmers who enter the pesticide-free, no-spray, clean-food, residue-free markets. The establishment of organic standards has been a primary stimulus for change in U.S. food production.

Organic Does Not Equal Sustainable. Sustainability implies a goal of

---

"closed-system" farming, meaning that farms approach self-sufficiency and require little outside input. However, most organic farms are "open-system," purchasing inputs produced off-farm, like fertilizer (rock phosphate, kelp, and fish).

In fact, many West Coast organic producers are biased against using livestock-based inputs, a common sustainable practice. Some claim that green manures and small amounts of off-farm inputs grow products with longer shelf life, better flavor, and fewer pest problems than products produced using livestock waste for fertility.

Many organic producers wonder whether any farm system can ever be sustainable in the pure sense. After all, organic systems still require cultivation, soil management inputs, processing, shipping, trucks, air freight, and farmer's market parking lots, all of which use oil, not usually produced on farms.

Comparing Apples with Apples. Also important was the use of a *norming* procedure in Iowa, Montana, and North Dakota. This was necessary because each state has several agricultural subregions that differ in crops, livestock, and production practices. The norming procedure ensured that each subregion was adequately represented in our selection of sustainable farms.

In the three "normed" states, norming was accomplished by comparing the index scores of respondents with others in their own subregions, rather than with all farmers statewide. In Minnesota, norming was not performed, and this created a disproportional representation of dairy farms in the sustainable group.

Commitment. We partly identified sustainable farmers by their *commitment,* rather than by how many years they had been using sustainable practices. Some farmers may have reduced purchased inputs and adopted "positive" practices for a while, not from commitment to enduring change but from temporary financial expediency. For some farmers, sustainable practices are a short-term strategy, especially at times of depressed prices. Should economic conditions improve, such facile commitments to sustainable agriculture may be abandoned.

Priority of Practices. Another issue is the priority with which farmers apply sustainable and conventional practices. Many conventional farmers use some sustainable practices, and sustainable farmers may continue to use conventional practices such as chemical controls, especially as a last resort.

Different priorities may be placed on practices. Sustainable farmers rely first on tillage or biological controls for controlling pests, and will use chemicals as a last choice. Conventional farmers are likely to make chemical pest control their first choice, with tillage a second resort.

Phase II

In Phase II of the research, subgroups of conventional and sustainable farmers were selected from those initially chosen in Phase I. The Phase II respondents fell at polar ends of the sustainability continuum—that is, those selected were conspicuously either "sustainable" or "conventional." Selected for inclusion in Phase II were 512 operators, of whom 279 (54 percent) provided information in personal interviews and mail questionnaires (Appendix A, Table A-3). Approximately half were sustainable farmers and the other half conventional farmers.

In Minnesota, a low response to requests for interviews forced classification of some farms as "sustainable," although they actually fell closer to the center of the sustainability index than was true for other states. Thus, distinctions drawn in later chapters between conventional and sustainable farms in Minnesota are less likely to result from their differences on the sustainability index than was the case for the other three states.

Among the farms we classified as sustainable, some operators would be the first to acknowledge that, although their farms may be more sustainable than their neighbors', true sustainability remains a distant goal. The respondents generally see sustainability as a future, and perhaps evolving, goal to be achieved, rather than a position they have attained.

A primary purpose of this study was to compare two divergent types of farming operations, each relying on a different set of production practices. Differences between these two types were identified and described. However, the *statistical significance* of these differences is not reported in this volume. Some of the differences were found to be not statistically significant, which is in part attributable to the small sample size in the four states and the fact that variation *within* each farm type sometimes exceeded the variation *between* the two types.

We are at an early stage in our exploration of sustainable agriculture. Considering this fact, plus our small samples, we elected not to adhere solely to the yardstick of statistical significance. For purposes of this volume, we believed it is more important to highlight apparent trends and relations, especially those important to policy deliberations, and to identify topics that are worthy of future investigation. (Please see Sidebar 3-5, "Six Caveats About Socioeconomic Impact Assessment.")

SIDEBAR 3-5

Six Caveats About Socioeconomic Impact Assessment

Gordon L. Bultena

Eric O. Hoiberg

During the four-state survey, we became keenly aware of six problems that impede rigorous evaluation of the socioeconomic impacts of sustainable agriculture.

1. What Exactly Is Sustainable Agriculture? Depending on one's viewpoint, sustainability variously implies:

• Greater profitability, achieved through less use of purchased production inputs (fertilizer and pesticide).

• More positive practices, such as substituting more environmentally benign inputs for potentially harmful ones.

• Fundamental, holistic redesign of farming operations to emulate and use natural systems (for example, securing greater diversity through use of more crop and livestock enterprises, soil fertility management through greater nutrient recycling, and use of biological and/or mechanical pest controls.)

These alternative definitions of sustainable agriculture are well-described in MacRae et al. (1990).

How sustainability is defined and measured can profoundly influence study findings. In this research, we used multiple criteria to gauge sustainability, including not only farmers' use of fertilizers and chemicals, but their adoption of positive practices such as replacement of purchased products with on-farm inputs. Also important was diversification through use of more crops in rotations and inclusion of livestock enterprises. (The criteria used in each of the four states to measure sustainability are presented in Table 3-1.)

2. Can We Attribute Observed Effects Solely to Sustainable Practices? Or are we seeing the influence of other factors that may distinguish the two farming styles, such as financial resources, farming experience, age of operators, and their different farm-management skills? The possible influence of such external factors is not addressed in this analysis.

3. Are We Seeing Actual or Perceived Impacts? For example, farmers may believe they are obtaining important personal/familial benefits from adopting sustainable practices, even though these benefits may not be evident from objective measures. Perceived change is important, for it often contributes to improved psychological well-being (greater job satisfaction and reduced stress) and future behavioral change (prompting experimentation with new farming techniques). Our study tested both perceived and actual impacts of sustainable agriculture.

4. Is an Impact Influenced by Mediating Factors? Some sustainable farmers have superior management skills or financial status, which may enhance their ability to secure benefits from sustainable practices.

63

For example, early adopters may reap windfall profits that are unavailable to later adopters. Consequently, we tested not only for differences between the two farm types, but examined the expression of specific impacts within each type. This required attention both to "average" impact scores and to the range or diversity of these scores within farm types.

5. How Soon Do Sustainability's Impacts Become Observable? Many impacts of sustainable agriculture may not become evident for years or even decades, and may not appear until a threshold number of farmers have adopted them. For example, sustainability often requires years of problem-solving with nutrients, weeds, insects, and management before optimum economic benefit is attained.

6. How Do We Gauge Community Impacts? Few sustainable farmers populate a typical community, so documenting their impact on the local economy and social patterns is difficult. Some studies (not of sustainable agriculture) have compared two or more communities populated by different types of farmers (Goldschmidt 1978; Labao 1990). However, such comparison was not feasible here because sustainable farmers tend to be widely dispersed and have low representation in any given community. Instead, we compared social and economic community linkages of conventional and sustainable farmers. From the nature and strength of these linkages, we deduced implications for community well-being.

However, a major shortcoming of this approach is that it can overlook or underestimate local impacts that might accrue if more sustainable farmers were present. These impacts may include development of new storage facilities, creation of specialized markets, and establishment of new retail businesses and farm services. Also, the presence of a core group of sustainable farmers in a community may foster more supportive local values toward alternative farming.

Sidebar 7-1, "Shelby County, Iowa: Sustainable Agriculture and Flexible Agribusiness," describes changes in a community that resulted when more sustainable agricultural practices were adopted by local farmers.

FOR FURTHER INFORMATION
Bultena, Gordon L., and Eric O. Hoiberg. 1992. Farmers' perceptions of the personal benefits and costs of adopting sustainable agricultural practices. Impact Assessment Bulletin 10(2): 43–57.

Conclusion: Defining Sustainable Agriculture

In studying sustainable agriculture, we learned lessons which may help others. These include the importance of a multidimensional perspective on sustainability, and the importance of being attentive to motives that underlie farmers' desires to change agricultural practices.

The complexity of farming practices and underlying motives is frustrating to researchers who study sustainability. We addressed this complexity by incorporating three key dimensions in our measurement of sustainability: adoption of reduced inputs, use of positive practices, and commitment to the values and goals of increased sustainability. Although it was difficult to substantiate with these data, our research left us with the impression of a very complex, dynamic interaction among these dimensions of sustainable agriculture.

4

Impact of Sustainable Farming on the Structure of American Agriculture

Gordon L. Bultena
Eric O. Hoiberg
Jodi Dansingburg

We compared conventional and sustainable farms on several criteria, including size of operation (acres), ownership, assets, sales, and enterprise diversity (farm products produced). Some important differences were found that may have implications for the future structure of agriculture. We found fewer differences when we compared some characteristics of the farm operators and their families.

INTRODUCTION

The structure of American agriculture has changed dramatically in recent decades. Millions of farmers and their families have been displaced by new technologies, more efficient production methods, and financial pressures. Labor input in U.S. agriculture has declined substantially in recent decades. Using 1977 as a reference year (index = 100), farm labor input declined from 265 in 1950 to 75 in 1990 (Hallberg 1992). Farms have grown larger and more capital–intensive. The forces driving farmers from the land are expected to intensify as the industrialization of American agriculture proceeds (U.S. Congress 1986).

Although recent changes in farm structure are seen by many Americans as constituting progress, others are alarmed by the rapid demise of medium–sized family farms and the increased control by large agribusiness firms over food and fiber production. For many proponents of sustainable agriculture, the diffusion of alternative farming practices is seen not only as a means of protecting environmental quality and improving food safety, but also of helping to ensure the survival of family farms and farm families.

An important question in the debate over sustainable agriculture is how its increased popularity might alter the long–established trends of declining farm numbers and expanded scale of farming operations. This question is examined here by comparing some personal and farm enterprise characteristics of sustainable (SUST) and conventional (CONV) farm operators in four states.

EFFECTS OF FARM SIZE: IT DEPENDS ON HOW SUSTAINABILITY IS DEFINED

Will the diffusion of sustainable agricultural practices stabilize the number of farms, or even bring an increase? The answer remains unclear, depending partly on what is meant by sustainability.

If sustainability is equated solely with reduced use of chemical fertilizers and pesticides, then increased sustainability could have little or no effect on dwindling farm numbers. Ironically, it often is the larger, more capitalized farms that are reducing agrichemical use through more precise application of less fertilizer and pesticides (Lasley et al. 1990).

But if sustainability means more integrated, holistic farming practices, smaller farms may result, because such practices require greater labor and management (Chapter 5), thus constraining the acreage that is farmable by an individual family. In fact, about half of Iowa's CONV farmers believe they can adopt sustainable practices only if they reduce their acreage (Bultena et al. 1992).

The natural limit on farm size imposed by an individual's ability to manage only so much area has been captured with the "eyes–to–acres ratio" concept. Wes Jackson (1985) has suggested that, for each type of farm, there is ". . . a natural limit to the number of acres a person can farm well. When that is exceeded, management skill is replaced by purchased inputs requiring more capital, energy, and nonrenewable resources."

Relation Between Farm Type and Size

We examined the relation between farm type and size by comparing the average acreage of conventional and sustainable farms in each of the four study states (Figure 4-1). The evidence discloses two important facts:

1. *Conventional farms commonly are larger than sustainable farms, often substantially so.* Average conventional farm acreage exceeds

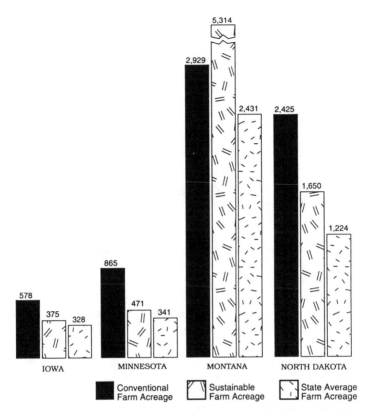

Figure 4-1. Farm acreage: conventional, sustainable, and state average (1991). State averages from USDA, *Economic Indicators of the Farm Sector: State Financial Summary, 1991.*

that of sustainable farms by 54 percent in Iowa, 84 percent in Minnesota, and 47 percent in North Dakota. (Montana is the exception, where sustainable farms/ranches are 81 percent larger. But the inclusion of rangeland in Montana contributed to much greater variation within both sustainable and conventional categories than was found between the categories, making this kind of comparison less meaningful than for other states.)

2. *Farm size varies substantially within both farm types.* For example, sustainable farms range from a few hundred acres to a thousand

acres or more in each of the four states. This wide variation suggests that sustainability, beyond the mere reduction of purchased inputs, may be feasible for some large operations.

It is noteworthy that each group of conventional and sustainable farms in this study *had an average farm size larger than the average farm size in its corresponding state.* One reason was that farms smaller than 50 acres were excluded from this study in Iowa, Minnesota, and Montana. Also, farms with annual sales of agricultural products less than $2,500 were excluded in the North Dakota study. Because our analysis excluded these small operations, farm-size indicators (acreage, assets, and gross income) for all eight groups could be expected to be larger than state averages.

We asked respondents, "Do you intend to increase your acreage in the next five years?" The most expansion-minded were CONV farmers in Iowa (37 percent) and North Dakota (55 percent) who planned to rent or purchase additional land. Fewer SUST farmers had expansion plans (20 percent in Iowa and 32 percent in North Dakota).

In Minnesota and Montana, however, equivalent proportions of CONV and SUST farmers reported expansion plans—approximately two-fifths of each group in Minnesota and one-fifth of each group in Montana.

Some scholars have suggested that the relation between farm type and size may vary regionally. For example, widespread sustainable adoption in the Corn Belt is predicted to halt, or even reverse, the historic expansion of farm size, whereas the effect could be the opposite in the Northern Great Plains. In the Plains, increased sustainability is seen as possibly encouraging even larger wheat and livestock operations (Dobbs 1993; Bird 1992).

ECONOMIC FACTORS

Sustainable agriculture often is claimed to have important consequences for ownership, financial assets, and farm sales. We examine each below. Chapter 6 compares the economic performance of the two farm types.

Ownership Status

The prominence of large, corporate farms in some states (especially California) concerns alternative agriculture proponents. They believe that everyone would be better-served by medium-sized, family-owned farms (Beus and Dunlap 1990). The 1987 Census of Agriculture found that most farms in our four-state study area are owned by individuals or families, about a tenth by legal partnerships, and a few by family corporations.

Comparable ownership patterns were found for the conventional and sustainable farms we studied. A substantial majority of both are owned by their operators (ranging between 63 and 90 percent across the four states). However, about one-third of Montana's conventional and sustainable farms are owned by legal partnerships or family corporations. In North Dakota, the same is true for about one-fourth of the conventional farms.

The 1987 Census of Agriculture found that a substantial portion of the acreage farmed in the four states is not owned by the farm operators:

- Only about half of the farm acreage in Iowa, Minnesota, and Montana is owned by persons farming the land; in North Dakota, it is one-third.
- Part-ownership is common, accounting for about a third of all farms in Iowa, Minnesota, and Montana, and half in North Dakota.
- Farmers not owning any of their land (tenants) make up about a fifth or less of all farms in each state.

Our study found that, on average, CONV farmers *own more acreage* than SUST farmers, which might be expected from their generally larger operations. But SUST farmers *own a larger proportion of their farmland* in Iowa, Minnesota, and North Dakota. This may reflect a greater value placed by SUST farmers on land ownership, the high capital requirements of full ownership of large operations, or both.

This larger average proportion of ownership by SUST farmers is important. First, it reflects the Jeffersonian ideal of farmers owning their own land. Second, widespread distribution of productive assets and wealth among farm operators is deemed important to competition and democracy. Third, maintaining integrated ownership and labor counteracts the trend toward absentee land ownership (ownership by those who may not care about the land) and large-scale corporate control of food production.

Financial Assets

Marked differences exist among the four states in the average farm asset value (land, buildings, machinery, crops in storage, livestock, supplies). Average asset values range from approximately $397,000 in Minnesota to $851,000 in Montana. In general, greater assets correlate with larger acreage (compare Figures 4-1 and 4-2).

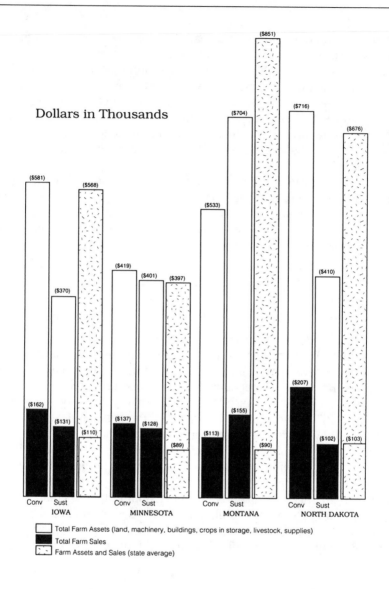

Figure 4-2. Farm assets and total farm sales of
conventional and sustainable farms (1991).

Conventional farms typically have greater assets, not surprising because of their larger acreage. For example, average assets of Iowa conventional farms are 57 percent greater than sustainables, 4 percent greater in Minnesota, and 75 percent greater in North Dakota. In Montana, where sustainable farms/ranches are larger than conventional units, the pattern is reversed, with sustainable farmers/ranchers reporting 32 percent greater assets (Figure 4-2).

Farm Sales

Conventional farms should have larger gross farm sales than sustainable farms due to their generally greater size. This was confirmed in Iowa, Minnesota, and North Dakota (Figure 4-2). In fact, 1991 average conventional farm sales exceeded sustainable farms by 24 percent in Iowa, 7 percent in Minnesota, and 103 percent in North Dakota. Sales by sustainable farms/ranches in Montana exceed conventional operations by 37 percent.

Average sales of farm products for the conventional and sustainable farms in our study were larger than corresponding state averages, except in North Dakota (Figure 4-2).

FARM ENTERPRISES FOR SUSTAINABLES FARMS

A *farm enterprise* is any product that a farmer elects to produce, either crop or livestock: beef cattle, corn, millet, raspberries, hogs, honey, and so on. We expected a larger number of enterprises on sustainable farms, and this was borne out. Our study examined two general types of enterprises, crops and livestock.

Cropping Enterprises

The amount and use of farm acreage varies markedly across Iowa, Minnesota, Montana, and North Dakota (1987 Census of Agriculture). Montana has the largest total land in farms/ranches (about 60 million acres) and Minnesota the least (about 27 million acres). Substantial proportions of farmland are in pasture and rangeland in Montana (68 percent) and North Dakota (29 percent). In contrast, harvested crops dominate farmland in Iowa (65 percent) and Minnesota (63 percent).

Major differences also occur in the states' principal crops (Figure 4-3). Only two (corn and soybeans) are planted on over four-fifths of Iowa's harvested cropland and over half of Minnesota's. Hay and oats occupy much smaller proportions of cropland in both states. Minnesota's climate permits adding wheat to the mix. As expected, North Dakota and Montana have a very different crop mix. Their cultivated land is about half in wheat, with significant acreages in barley and hay.

73

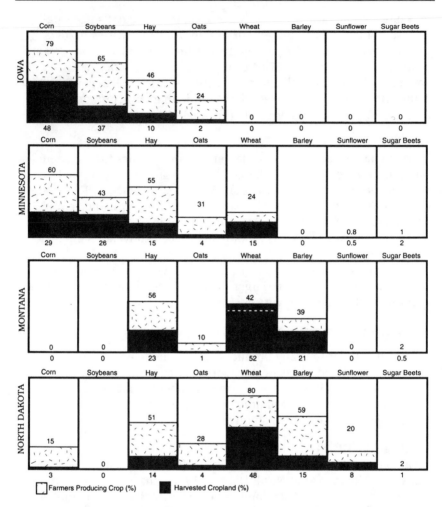

Figure 4-3. Proportion of all farmers producing each crop and proportion of harvested cropland by state (1987 Census of Agriculture).

By definition, sustainable farms produce more varied crops than conventional farms. For example, we found that Iowa's sustainable farms have nearly twice the number of crops on average (4.0) than are grown on conventional farms (2.4).

The greater cropping diversity of sustainable farms is further demonstrated by the proportion of the two farm types that produce specific crops. Crops grown on a fifth or more of either conventional or sustainable farms in each state are shown in Figure 4-4. Among the states, cropping diversity is greatest in Montana and North Dakota, but in all four states, it is greatest on sustainable farms.

Not only do SUST farmers produce more varied crops; they also seem more committed to future diversification. When asked, "Do you plan to add new crops in the next five years?" more SUST than CONV farmers said "yes" in Iowa (29 versus 10 percent), Montana (46/17 percent), and North Dakota (51/37 percent). In Minnesota, the reverse was true; more CONV farmers planned new crops (37/27 percent).

Farmers who raise fruit and vegetables have different opportunities and risks than those producing grain or livestock. Please see Sidebar 4-1, "Horticultural Farming—A Different Ball Game."

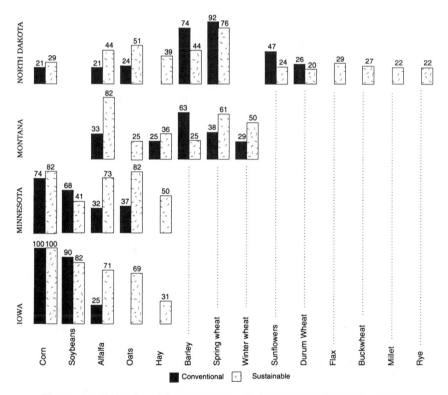

Figure 4-4. Diversity of Conventional and Sustainable Cropping, 1991. Numbers are % of all farms producing the crops. Crops are grown on at least one-fifth of farms in each state.

SIDEBAR 4-1

Horticultural Farming—A Different Ball Game

Harry MacCormack

Horticultural farmers, who raise fruit and vegetables, have different opportunities and risks than grain or livestock farmers. Horticultural farming can be accomplished on very small acreages. In Oregon's Willamette River Valley, a family may support itself with three to ten acres of horticultural crops, bringing a gross income of $5,000 to $10,000 per acre. Families have grossed $250,000 or more on just four acres.

Horticultural farming is niche- and market-oriented, much in specialty crops like "baby" vegetables (baby carrots, baby limas, and early-harvest lettuce); high-quality blueberries, raspberries, and blackberries; nuts, such as filberts; and vegetable seeds. These specialties allow very small farms to survive while many larger farms are in financial trouble.

Horticultural farming is labor-intensive. Labor often must be hired from outside the farm family or community, often for only a day or a few weeks during cultivation or harvesting. Horticultural farming also is highly market-driven and closely attuned to consumer fads. Farmers in this agricultural sector often are intensively entrepreneurial in their business management.

Horticultural farms qualify for little government support. To encourage these small but important enterprises, direct support (by the farm program) and indirect support (through education and extension) are needed.

Livestock Enterprises

Historically, livestock have been integral to farming in these four states. But increased specialization and the intense labor that livestock demand have sharply reduced the number of farms with livestock.

As expected, sustainable farms are more likely than conventionals to have livestock operations (Figure 4-5). Most SUST farmers in Iowa, Minnesota, and Montana raise livestock (93, 91, and 89 percent), compared to smaller proportions of CONV farmers (46, 37, 58 percent). The same pattern holds for North Dakota, but is less pronounced. In addition to their greater likelihood of having livestock enterprises, a larger number of sustainable farmers than conventional farmers in Iowa, Montana, and North Dakota had multiple livestock enterprises. For example, from a third to a half of the sustainable farmers with livestock in these three states reported three or more different livestock enterprises, compared to only about a tenth of the conventional farmers. Although more SUST farmers have livestock, they differ little from conventionals in production methods. This similarity may reflect many SUST farmers' initial concentration on attaining greater sustainability

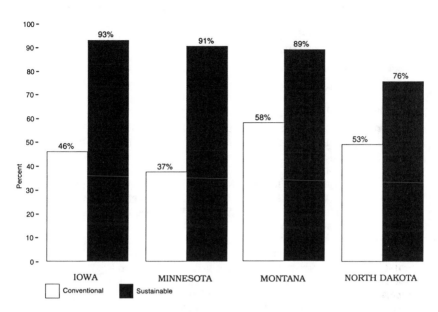

Figure 4-5. Proportion of conventional and sustainable farms having livestock enterprises, 1991.

in crop and animal manure management, thus delaying efforts to make their livestock enterprises more sustainable.

In fact, Minnesota SUST farmers often reported modifying livestock practices only after having solved nutrient and weed problems. Also, until recently, sustainability research has focused primarily on crop and manure management, not on livestock production.

A marked exception is the recent popularity of *intensively managed rotational grazing*. This technique is being adopted by growing numbers of beef, dairy, and other livestock producers (Sidebar 4-2).

SIDEBAR 4-2

Controlled Grazing

Jodi Dansingburg

Charlene Chan-Muehlbauer

Douglas Gunnink

Controlled grazing (also called "rotational grazing" or "management-intensive grazing") is a sustainable practice that is rapidly replacing traditional confinement dairying (also called "row-crop-and-feedlot," or "continuous grazing").

Used most often to manage dairy, beef, and sheep herds, controlled grazing also is used for swine and poultry. The practice requires much lower investment in buildings, manure handling, feed storage, and equipment. It is praised by adopters for improving profitability, impacting the environment less, and improving quality of life by reducing labor.

Managing Plant Growth Is the Key. Controlled grazing begins by establishing a diverse pasture (mixed grasses and legumes). The pasture is divided into "paddocks" with special electric fencing, using lightweight fiberglass fenceposts and flexible fencing wire. Animals are systematically cycled through the paddocks to obtain optimum forage and to give plants time to regrow. Controlled grazing requires a high level of pasture and animal management.

Case Study: The French Family Dairy Farm

The French family operates a 325-acre dairy farm in southeastern Minnesota. Prior to 1989, their conventional farm's production and profit were average-to-above for their area. But, despite sound management and solid performance, the farm was not improving financially.

Following the difficult drought of 1988, the Frenches adopted controlled grazing and focused on increasing their net income, rather than their gross. Controlled grazing let them drastically reduce feed, fuel, electricity, labor, and bedding cost, saving $12,356 in 1989, $22,124 in 1990, and $15,685 in 1991.

The average 1991 net farm income for southeastern Minnesota farmers was $29,231, and the French's savings were more than half of that, clearly translating to a dramatic income increase. In fact, their net "return" of money to labor and management more than doubled, from $14,413 (1988) to $37,863 (1991). (Average 1991 income from 1992 Annual Farm Business Management Report, Southeastern Minnesota, April 1993, Minnesota Riverland Technical College, Austin Campus.)

More Benefits than Money. The Frenches converted much of their tillable land to pasture, significantly reducing soil erosion. They also eliminated herbicide and fertilizer, so fear of chemical run-off or leaching is gone. Herbicides are unnecessary on their pastures because most crop weeds make nutritious forage. Manure is spread evenly through the paddocks by the livestock themselves, adding fertility yet avoiding excess accumulation and run-off. Earthworms, essential to soil quality, have increased substantially.

Cattle health also has improved since the Frenches began controlled grazing. "Our grazing system lets the cattle follow their natural instincts," Dan comments, "As a result of the reduced stress on the animals, they view us as a friend instead of an enemy." The Frenches believe it now will be possible to keep cows in their herd for ten years, a dramatic increase over the industry average of 2.4 years.

With their cows mostly on pasture, daily feeding and manure management are reduced. Less time is spent producing feed. The reduced workload also has reduced the worry and effort of planting and harvesting.

Increasing Their Options. Like many others who have adopted controlled grazing, the French family discovered they have more land than needed for their herd. They adjusted by adding more cows, providing additional income to pay debt. Later, they may cut their herd to increase leisure time, or rent their excess land to a beginner who could use their milking facilities and start building a profitable herd.

Potential Impact on the Dairy Industry. As of mid-1993, Minnesota's 13,500 dairy farms were going out of business at the average rate of 15 farms per week, according to the Minnesota Department of Agriculture. That is nearly 800 farms per year. Departing young dairy farmers most often cite lack of profitability and excessive workload as reasons for leaving.

Controlled grazing is an unusual new technology, because it relies on much less mechanization than the technology it replaces. It also shifts some agricultural activity back onto the farm. Because it is simpler and less expensive than confinement housing, it is easier to adopt by farmers who are beginners or are smaller and less-capitalized. Thus, controlled grazing offers family-size farmers a chance to reduce workload and increase profitability.

Charles Opitz, a Wisconsin dairy farmer who manages his 1,000-plus dairy herd with controlled grazing, observed: "If I could start over again with controlled grazing, I would milk 120 cows with no hired help on a seasonal basis and make over $100,000 net income."

FOR FURTHER INFORMATION

Land Stewardship Project. 1994. *An agriculture that makes sense: Profitability of four sustainable farms in Minnesota.* A 12–minute video available from Land Stewardship Project, P. O. Box 130, Lewiston, Minnesota 55952, 507/523–3366.

PERSONAL AND HOUSEHOLD CHARACTERISTICS

Personal Characteristics

In addition to farm enterprise characteristics, we tested for possible differences in some personal and household characteristics of the two farm types. Characteristics such as age and educational attainment can have important effects upon the receptivity of farmers to adopting new agricultural practices, and the success with which these practices are implemented. Familial characteristics, such as the number of children at home, can be important for the greater farm labor often associated with sustainable agriculture.

Age differences of farmers between the two farm types are small. The average age of farmers in each state for both types is mid-to-late forties (the span is 20 to 83). (In North Dakota, farmers over the age of 65 were excluded from the study.) However, these are the respondents' *current* ages, and not their ages when they first began implementing alternative practices. For many, this occurred considerably earlier in their farming careers—see Chapter 9.

Educational attainment of the two farm types also is quite similar, although these patterns differ by state. In Iowa, CONV and SUST farmers have roughly equivalent levels of educational attainment, with about half in each category having completed some post-high school education. However, CONV farmers in North Dakota had the higher levels of educational attainment, with three-fifths having completed some schooling beyond high school compared to about two-fifths of the SUST farmers. In Minnesota, four-fifths of the CONV farmers, and less than half of the SUST farmers, report formal schooling beyond high school.

Gender. Consistent with national patterns, virtually all of the CONV and SUST farm operators are male. However, as shown in Chapter 5, spouses in both types of operations often make important contributions to farm management and labor.

Household Characteristics

Household characteristics of the two farm types also are similar in all four states. Average household size (number of persons living at home) is three to four persons for both farm types. A spouse is present in nearly all households.

A sizeable number of the households (about two-fifths) no longer have children at home, partly reflecting the older age of many respondents. Households with children usually have two or three, with their ages averaging from 8 to 17. In three of the states (Iowa, Montana, and North Dakota), SUST farmers are the most likely to have children at home.

Labor-intensive practices common on sustainable farms often require labor from spouses and children (Chapter 5). However, as farmers and spouses age, and children leave home, farm management may change or hired labor may be needed.

Are sustainable farms more likely than conventionals to be farmed by family members upon retirement of their operators? To find out, we asked respondents, "What will happen to your farm when you retire? Will it be farmed by family members or relatives, rented or sold outside the family, or what?" By small margins across the four states, more SUST farmers than CONV farmers believe their farms will continue operation under family members or other relatives. More CONV farmers often anticipate their farmland being sold or rented outside their families.

CONCLUSION

How sustainable agriculture will impact the structure of American agriculture is highly dependent on how sustainability is defined. If it is equated solely with reduced chemical use, sustainable farms in the future could be large–sized operations, and a continued decline in farm numbers and population would be expected. However, if sustainability implies the substitution of "positive" nutrient and pest control practices for purchased production inputs, then smaller farms should prevail.

Our finding that sustainable farms typically are smaller in size (acreage) than conventional farms suggests that widespread adoption of sustainable farming practices should help retard the historic decline in farm numbers and population, especially in the Corn Belt. However, the likely effects of sustainable agriculture on farm numbers may not be as dramatic as sometimes portrayed. Moreover, sustainable farms range widely in size. Optimal farm size under different sustainable crop and livestock production systems remains unclear.

Conventional farms not only tend to have greater acreage than sustainable farms, but tend to have larger financial assets and sales of farm products as well. (See Chapter 6 for a comparison of their financial performance.) The effect of sustainable agriculture on farm numbers also will be affected by the ability of sustainable farms to maintain financial viability.

By definition, sustainable farms are more diversified than conventional farms. They commonly produce a greater variety of crops, practice more complex crop rotations, and more frequently have one or more livestock enterprises. An increase in the number of such farms could have important implications for crop and livestock markets as well as impacts on the land.

The advanced ages of many of these conventional and sustainable farmers suggests that our society is entering a "watershed period" in which the future disposition of their farmland will be critical to the success of sustainable agriculture. For sustainable agriculture to increase, it is imperative that this farmland not be consolidated into larger units, but instead be made available to younger farmers, who are the most inclined to adopt sustainable agricultural practices (see Chapter 9).

5

Farm Labor and Management

Keith Jamtgaard

Both conventional and sustainable farmers perceive that sustainable practices require more labor. In fact, sustainable farmers work substantially more hours per year than conventional farmers in three of our four study states. Labor demand of sustainable farms is more year-round and less seasonal. This greater labor requirement for sustainable farms could help limit their size and thus create more employment opportunities in farming.

INTRODUCTION

When farmers or ranchers consider adopting new farming practices, their usual concerns are the labor required and its likely financial return. Both conventional (CONV) and sustainable (SUST) farmers believe that alternative farming practices require more labor and managerial skill (Bultena et al. 1992).

Some farmers view these demands positively because of their potential for increasing profitability and the often welcome psychological challenges and rewards of farming more sustainably. Increased labor could improve both income and opportunities on farms that earn a return on this additional labor.

If SUST farmers net a larger share of gross farm income—for example, through use of family labor to replace purchased inputs—they can earn higher returns on the capital they have invested in their farms. If widely adopted, farming systems that use more family labor to capture a larger share of the farm dollar can be expected to slow the growth in farm size and stabilize the number of farmers, or at least slow the present decline.

For some farmers, however, the poential for increased labor

demand poses a barrier to adopting sustainable practices. Their concerns are how to provide the needed labor, the greater time demanded, and whether more intensive labor is the best approach to improving profit.

For most agricultural production, labor demand is uneven. The seasonal nature of farming often requires a fixed sequence of activities. This presents a dual problem: obtaining sufficient help during busy periods, and keeping employees productive during waiting periods between stages of crop production (Pfeffer 1983). Labor-saving technological advances have exacerbated this labor problem, both by enabling farm expansion and by making the labor demand more uneven (Mann 1990; Pugliese 1991).

Agriculture's uneven labor demand creates an opportunity for alternative cropping and livestock enterprises that increase the time farmers spend creating value in their operations during otherwise slow periods. On the other hand, use of additional labor to generate on-farm inputs and to service livestock can unintentionally aggravate existing labor and management problems.

The relative labor demand of conventional and sustainable farming remains largely undocumented. Nor is there much information about the financial return from the added labor required by sustainable farms. In this study, we examined the amount and source of labor on both farm types (reported in this chapter), and assessed the financial return from this labor (Chapter 6).

One of the most straightforward yardsticks of farm labor is time spent in farm-related activities, so we measured that. But even this seemingly simple measure can be viewed in several ways: amount of labor time, who provides this labor, labor variation across seasons, work role allocations, activities accomplished by workers, who makes farm-management decisions, and whether farm work is coupled with off-farm work. We also examined respondents' perceptions of the skill required for successful operation of conventional and sustainable farming practices.

TIME REQUIREMENTS OF FARM LABOR

We estimated time spent on farm-related work (in 1991) for three labor sources: household members, contracted or custom services, and other labor sources outside the household. Except for contracted and custom services, we obtained data on the relationships between labor providers and farm operators, demographics of the workers, their ownership interest in farm operations, and time spent working on farms during several seasons of the year.

In three of the four states, SUST farmers in our sample spent more hours per week in farm labor than CONV farmers (averaged annually). Viewed over the entire year, farm-related work reported by SUST farmers was 35 percent greater than CONV farmers in Minnesota, 23 percent greater in Iowa, and 19 percent greater in Montana (Figure 5-1a). No difference was found by farm type in North Dakota.

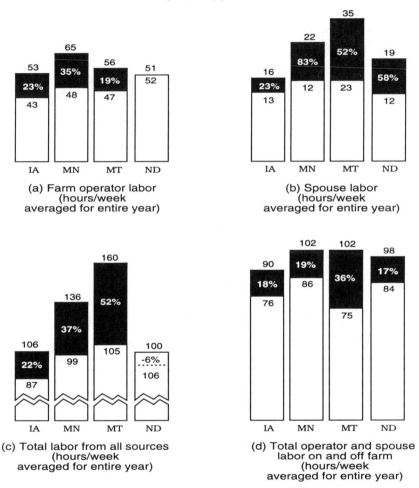

(a) Farm operator labor
(hours/week
averaged for entire year)

(b) Spouse labor
(hours/week
averaged for entire year)

(c) Total labor from all sources
(hours/week
averaged for entire year)

(d) Total operator and spouse
labor on and off farm
(hours/week
averaged for entire year)

Figure 5-1. # Hours of Farm Labor (1991). Upper numbers show sustainable hours; lower numbers are conventional hours. Percentages show how much greater sustainable hours are than conventional.

Difference in hours worked by the two farm types was even more dramatic when comparing the labor contributed by spouses (Figure 5-1b). Among spouses on Minnesota sustainable farms, the average for the entire year was 83 percent higher than for conventional operations. Spousal labor on sustainable farms was 58 percent greater in North Dakota, 52 percent in Montana, and 23 percent in Iowa.

Farm labor time contributed by *all* labor sources—household members, contracted and custom services (for example, fertilizer application or custom harvesting), and other sources outside the farm household—was substantially greater for sustainable farms than for conventionals in Iowa, Minnesota, and Montana (Figure 5-1c). North Dakota was the exception, with conventional farms reporting the greater labor investment.

On average, Minnesota and Iowa sustainable farms, although smaller than conventionals, required greater labor. North Dakota sustainable farms were substantially smaller than the conventionals but required nearly as much labor. It follows that the labor invested *per acre* on these sustainable farms also was greater. Montana operations were an exception because, on average, sustainable operations were larger than conventionals, primarily due to more range and pasture, requiring less total labor input per acre.

One explanation for sustainable farms' generally greater labor is their greater use of livestock. We observed a pronounced pattern of greater labor with livestock operations on both types of farms.

SEASONAL DIFFERENCES IN FARM LABOR

Because farm labor is seasonal, it is unevenly distributed through the year. In some enterprises—for example, vegetable production—labor can be highly concentrated in a particular season, so that the farm family may be unable to provide all the labor, thus requiring temporary outside help.

In contrast, a more even seasonal distribution of labor is conducive to family operation of farms, or the use of permanent hired labor, rather than relying on temporary outside help. Comparing the summer and winter graphs in Figure 5-2 discloses that labor is more evenly distributed through the year on sustainable farms than on conventional farms.

Figure 5-2 also shows the average winter work hours per week as a percentage of the average summer work hours per week. CONV farmers worked 44 percent of their summer workloads in Minnesota, 48

86

percent in North Dakota, 55 percent in Montana, and 59 percent in Iowa. In contrast, SUST farmers worked 67 percent of their summer workloads in Montana, 58 percent in North Dakota, and 71 percent in Iowa and Minnesota. Similar patterns were found for spouses and for households as a whole.

These results show that, although sustainable farms require more labor, they do so in a more balanced, year-round manner than conventional farms. The critical question--how adequately SUST farmers are financially compensated for this greater labor input--is addressed in Chapter 6.

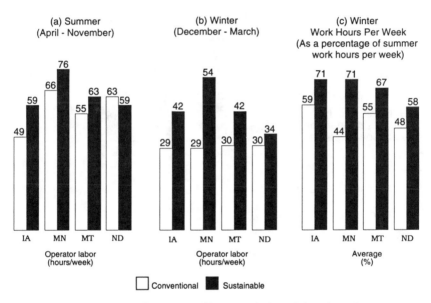

Figure 5-2. Seasonal differences in farm labor (1991).

OFF-FARM WORK

Off-farm employment of farmer and spouse is common. It sometimes reflects the inability of a farm to provide adequate income for the farm family, and to provide full employment for family members. Because sustainable practices generally require more labor, employment of family members off-farm may limit the labor available to the farm, and thus limit adoption of sustainable practices. At the very least, off-farm employment may lessen the farm family's ability to effectively implement sustainable practices.

We asked respondents whether they and/or their spouses were employed off-farm during 1991, and the hours worked per week. In Iowa and Minnesota, SUST farmers worked off-farm less often than conventionals. For spouses, the percentage reporting off-farm work was greater for SUST than CONV farms in Iowa and Montana, but this dif-

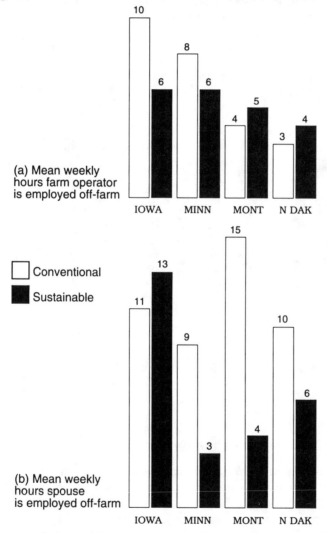

Figure 5-3. Off-farm employment of farmers and spouses (1991).

ference was reversed in the other two states. However, in Montana and North Dakota, SUST farmers were more likely to work off-farm. Overall, conventional farmers averaged more hours of off-farm work in Iowa and Minnesota, but sustainable farmers had more hours in Montana and North Dakota. Spouses of CONV farmers averaged more off-farm work hours than their SUST counterparts in three of the four states, with Iowa being the lone exception (Figure 5-3).

Spouses from sustainable farms were more likely to hold off-farm jobs in Iowa and Montana, but less likely in Minnesota and North Dakota (Figure 5-3b). Total off-farm hours worked by both farmers and spouses revealed substantial differences only in Minnesota and Montana.

In Montana, SUST farmers and spouses worked over twice the hours off-farm as conventionals, but the opposite was true for Minnesota. Iowa and North Dakota also showed a pattern of less, or nearly equal, off-farm work among sustainable respondents, compared to conventionals. These findings clearly run counter to the stereotype of SUST farmers as part-time farmers who work primarily off the farm.

Because of the large amount of farm labor, cumulative work time in all states (off-farm plus on-farm work for both spouses) clearly was greater for sustainable farms (Figure 5-1d).

WHO PROVIDES FARM LABOR?

The percentage of farm labor performed by household sources was 72 to 91 percent of the total workload on farms of both types. No clear difference in this measure was found between the CONV and SUST samples. However, due to the larger number of hours worked on sustainable farms in three of the four states, household sources contributed more hours in these states. Farmers account for most of this workload, but spouses and family members also contribute significant labor.

Contract or custom labor, although clearly minor, comprises a slightly larger share of conventional farm labor than sustainable, except in Minnesota.

We also examined non-household labor, apart from custom or contract services. This labor pool, including neighbors, friends, and non-resident hired help, contributed 9 to 27 percent of all farm labor across the four states.

In three of the four states (excluding North Dakota), non-household labor accounted for a larger labor share for sustainable operations, compared to conventionals. However, with the exception of Iowa, the outside labor on sustainable farms was much more likely to have an ownership interest in the operation than was the case for conventional operations.

Farm Production Versus Farm Business Tasks

We looked at the allocation of responsibility for performing farm tasks in two categories: farm production and farm business. *Farm production tasks* consist of fertilizer application, weed control, care of livestock and pastures, planting, and tillage. *Farm business tasks* encompass farm finances, bookkeeping, and purchase of inputs. We asked respondents to identify the specific persons having primary responsibility for each task.

Farm Production Tasks. Generally, CONV farmers were equally as likely as SUST farmers to have sole responsibility for production tasks. Only in Montana did sustainable farm families have greater involvement in production (that is, the farmer performed fewer tasks alone). In Iowa, these tasks most often were confined to the farmers themselves.

Farm Business Tasks. In our sample, SUST and CONV farmers differed significantly in only one farm business task—handling finances. In three states, SUST farmers were more likely to handle farm finances, whereas CONV farmers more often shared this task with others—a partner, professional, spouse, or child. In Minnesota, the exception, SUST farmers handled finances less often by themselves. Farm bookkeeping for both conventional and sustainable operations was more frequently a collaboration with others, often a spouse, child, or professional. In fact, a spouse not infrequently bears sole responsibility for this task.

FARM DECISION-MAKING

We studied several types of farm activities to determine whether farm decision-making varies by farm type. Respondents were asked to name the primary decision-maker for each listed activity. Decision-making patterns differ both between states and farm types. Sustainable households in Minnesota, Montana, and North Dakota typically had more shared responsibility than conventional households, where farmers more often made decisions alone. No clear differences in decision-making patterns were found between CONV and SUST farmers in Iowa.

SKILL REQUIREMENTS

Skills required for 17 farm activities were assessed for the two farm types. Respondents scored each activity from 1 (little or no skill required) to 10 (a great deal of skill required). We found little difference in these skill scores of CONV and SUST farmers. The only activity for which SUST farmers in all four states perceived the need for substantially greater skill was animal manure management. This is not surprising, because more sustainable farms include livestock in their production systems, and the efficient and effective use of manure is a goal for many of these systems.

However, subtle differences exist in some of the skill ratings. Several tasks in finance and marketing were seen by both groups as generally having somewhat higher skill requirements than tasks in management of hired labor or farm planning. But CONV farmers were more apt to consider financial and marketing tasks as requiring greater skills than were SUST farmers. SUST farmers attached somewhat higher skill scores to labor and planning activities than did CONV operators.

MANAGEMENT COMPLEXITY

A common perception, even among many sustainable agriculture proponents, is that its adoption increases management complexity. To explore this, we asked two questions about farm management problems:

1. *What activities are most difficult to complete on schedule?*

2. *What two or three farming-related activities pose the greatest challenge because they require new knowledge, techniques, or skills?*

Both CONV and SUST farmers in all four states concurred that *cropping* and *soil practices* were the most difficult to complete on schedule. No other activities were seen as presenting serious problems of timely completion.

CONV farmers in all four states most often mentioned *farm product marketing* as their greatest challenge. This includes selling grains, securing good prices, grain contracting, and livestock marketing. The reason they considered marketing most challenging was because it required new knowledge. Marketing also ranked among the top three concerns of SUST farmers in all four states, especially the marketing of specialty and organic crops.

SUST farmers in three of the four states listed their most difficult task as *managing crop and soil practices*. Specific concerns were many, including new tillage practices (no-till, ridge tillage, conservation tillage), soil fertility (which fertilizer to use, soil nutrients, soil sampling and testing, choosing soil inputs), problems in maintaining organic methods and certification, keeping up with new technologies, choosing seeds, soil conservation (terracing, contouring), diversifying crops, incorporating new crop rotations, dealing with problems of continuous cropping, and managing soil moisture.

Not surprisingly, livestock management was among the top three concerns of SUST farmers in three of the four states (Minnesota excepted). This included all aspects of livestock production (purchasing, nutrition, animal health, breeding, calving, genetics), improved animal productivity, use of new pasture-management techniques (for example, intensive short-term rotational grazing), improvement of dairy practices, and staying current with new milking techniques.

These differences in management challenges may be important for the structure of agriculture. Marketing, the biggest challenge to CONV farmers, is more easily separated from farm operation than production problems, which are cited most often by SUST farmers.

The more production is routinized, the fewer people are needed to exercise judgment and management, and therefore the more feasible it becomes to separate farm operations from ownership and management. This could facilitate a shift away from the family farm system, toward one in which labor is largely unskilled. To the extent that sustainable farms present greater management challenges in the field and barn, it may be more difficult for farm owners to relegate production tasks to unskilled workers.

CONCLUSION

The perception that greater farm labor accompanies a shift to a more sustainable agriculture generally was supported in three of the four states. An important reason was the more common livestock enterprises of SUST operators, which require more labor. Labor use generally was distributed more evenly through the year on sustainable farms, whereas conventional farms showed larger differences between summer and winter workloads.

Overall, the greater labor demand on sustainable farms gives them the potential to increase rural economic opportunities, both for farm household members and residents of local communities. However, these labor opportunities can be translated to real economic opportunities only if farming is paying adequate return for that labor. This is the subject of the following chapter.

6

Economic Position and Performance of Sustainable Farms

David L. Watt

The sustainable farms had greater debt and faced more precarious financial circumstances than conventional farms in 1991. However, there is considerable variation within both groups. Although some are doing well, the average financial performance of both types in 1991 was insufficient to pay for family labor and capital invested in operations, while maintaining net worth.

Farming is a business. It is a very difficult one, dependent on the weather, market, and the farmer's knowledge and skill. In few other enterprises are efficiency and productivity so critical. Farmers are self-employed people who must do the work, manage the business, provide their own paychecks and benefits, and show a profit. Farming is financially risky, with great year-to-year variation in farm income.

Farming is hard labor, yet the farming business itself is fragile. A farmer must be very careful in adopting new practices (like reduced chemical use) and buying capital equipment (a new $70,000 tractor, for example). A miscalculation can devastate profit and put the farmer out of business. Because farming is a business, any farm must remain profitable if it is to survive.

Thus, any decision to adopt sustainable practices must be strongly influenced by the *economic wisdom of doing so.*[1] Unfortunately, it is difficult for farmers to assess the true economy of such adoptions, for there is little dependable data on the economic performance of alternative farming practices (National Research Council 1989, 195). Not surprisingly, "this lack of information may be a major deterrent to [alternative] system adoption" (Ikerd et al. 1993, 38). In part, this lack of information prompted our study.

---

[1]The centrality of economics in choosing to adopt alternative farming is highlighted in Chapter 4 of the book *Alternative Agriculture*, by the National Research Council (1989).

Key questions that we asked were:
- *How do sustainable and conventional farms compare in financial position?*

- *Is our economic system providing sufficient economic incentives to encourage farming practices that support a healthy environment?*

Prior research clearly shows why it is so difficult to make simple statements about the economics of sustainable agriculture:

- *Alternative production systems vary significantly from year to year and by location, because they are site-specific and farm-manager-specific.* This conclusion is from a review of comparative economic studies of alternative farming systems through 1989 (Fox et al. 1991).

- *Most economic studies of low-input farming have been at the individual crop or livestock level; they show mixed results, with some potential.* This highlights the void of data for whole-farm analysis, which is needed to determine the financial position of sustainable farms (Madden and Dobbs 1990).

In Chapter 4, we compared the size of sustainable and conventional farms. In this chapter, we make economic comparisons using indicators of financial position and economic performance. Financial position is measured here by solvency and net worth, for they indicate the general condition of the whole farm. (For a summary of other comparative studies, please see Sidebar 6-1, "Other Studies Compare Conventional and Sustainable Profitability.")

SIDEBAR 6-1

Other Studies Compare Conventional and Sustainable Profitability

Thomas L. Dobbs

Studies Through the 1980s

If ecologically sustainable systems are not profitable to farmers, then we should restructure society to make them profitable, according to Crews, Mohler, and Power (1991). This would be better than declaring such systems unsustainable because they do not presently satisfy both ecological and economic criteria.

In that spirit, some economists have begun multidisciplinary research to compare the profitability of conventional farming with more ecologically sustainable systems. Some of this research has involved long-

---

term comparisons, in contrast to the single-year (1991) economic comparison described in the present work, and more direct and controlled comparisons than were possible here.

Sustainable systems sometimes are less profitable, but that does not mean they are inferior in the larger social picture. In fact, if some social costs are estimated in monetary terms (e.g., Faeth et al. 1991), the overall economic value of many sustainable systems may prove superior to their conventional counterparts. However, if farmers are to adopt ecologically sustainable systems *voluntarily,* public policy or other factors may require change to make these systems as profitable as conventional systems.

Cacek and Langner's 1986 literature review revealed mixed results regarding profitability. Profits were greater for conventional systems in some instances, and greater for organic systems in others, depending on the geographic area and farming enterprises. (Note: In most studies, organic farming is considered to be a variety of sustainable farming.)

Madden and Dobbs (1990) reviewed literature through the late 1980s and found Integrated Pest Management (IPM) economically promising. However, systems like IPM do not always decrease use of chemical pesticides. The authors concluded that emphasizing legumes in rotation and minimizing or eliminating synthetic chemical inputs both offer encouraging farm-level profitability prospects.

However, a concurrent literature review by Crosson and Ostrov (1990) reached more negative conclusions. Except for comparisons of organic and conventional Corn Belt farms in the 1970s by Lockeretz and associates (for example, Lockeretz, Shearer, and Kohl 1981), the studies they reviewed showed sustainable farms to be less profitable than conventional.

Fox et al. (1991) recently reviewed North American literature that compares the profitability of three farming systems: organic, other sustainable (which they call "alternative"), and conventional. The authors found that neither organic nor conventional farming has consistently outperformed the other in profitability.

Results also were mixed in studies that compared conventional systems with "alternative" farming (sustainable but not organic). Overall, the profitability findings depended not only on variations in production systems and crops produced, but also on weather, soil type, and assumptions about price and cost structure.

Newer Studies

In more recent studies, Ikerd, Monson, and Van Dyne (1993) examined nine U.S. resource areas and found opportunities for increased short-run profits with sustainable systems, compared to conventional systems. However, this study was unable to compare long-run profits.

Long-run profitability was compared in a northeastern Iowa study by Chase and Duffy (1991). They found a reduced-chemical corn-oats-alfalfa system to be as profitable as a conventional continuous corn system, but much less profitable than a conventional corn-soybean system.

Findings were similar in a long-term study (Dobbs and Henning 1993) on the western edge of the Corn Belt (east-central South Dakota). In comparing two operating farms during 1985-1992, the authors found respectable returns for a conventional corn-soybean farm and a largely organic small grain-alfalfa-soybean-corn farm. However, the conventional farm was considerably more profitable.

In Ohio, on the opposite side of the Corn Belt, Batte (1992) compared whole-farm profits for organic and conventional farms. The author estimated that the organic farms were more profitable, even though they operated fewer acres, but their greater profitability largely was due to price premiums for organically grown products.

Recent studies for the eastern U.S. indicate that policies or technology may need to evolve if eastern sustainable farms are to become as profitable as conventional. An economic analysis of long-term agronomic data (1981-1989) at the Rodale Research Farm in Pennsylvania compared a conventional farming system (corn and soybeans) to a low-chemical input system (corn, soybeans, small grain, and red clover), finding the conventional system more profitable (Hanson et al. 1990).

A Virginia study showed that two alternative farming systems using winter cover crops as green manure were less profitable initially than a conventional farming system (Norris and Shabman 1992). However, after a transition period, net returns of the alternative systems approached those of the conventional system.

Profitability Pattern

A pattern emerges from recent studies: sustainable systems at present appear more competitive with conventional systems in predominantly small-grain areas, or in transition areas between the Corn Belt and small-grain areas, than in the Corn Belt. Diebel, Llewelyn, and Williams (1993) recently compared a conventional system in northeastern Kansas with four different alternative systems using forage or green manure legumes in their rotations. They found two of the alternative systems more profitable than the conventional system, and two less profitable.

Two long-term studies (1986-1992) in northeastern South Dakota are reported by Dobbs (1994). In them, sustainable systems returned profits nearly equal to or greater than conventional systems, even disregarding premium prices for organic products. Other studies in wheat areas of South Dakota, based on case farms, also indicate that some sustainable systems may be competitive with conventional systems (Dobbs, Taylor, and Smolik 1992).

The National Research Council (1989, Chapter 4) identifies seven goals of farm economic performance:

Cost, Productivity, Profitability, and Capital Expenditures

1. Lowering per-unit expenditure on production inputs.
2. *Productivity*—increasing output per unit of input.

3. Producing more profitable crops and livestock.

4. Reducing capital expenditure on machinery, irrigation equipment, and buildings.

5. Making fuller use of available land, labor, and other resources.

Risk Measures

6. Reducing natural losses of crops and animals.

7. Reducing income loss caused by commodity price fluctuations.

Our economic performance comparisons in this chapter address the first five issues.

ABOUT OUR SURVEY

We examined the 1991 performance of a sampling of conventional farmers (CONV) and sustainable farmers (SUST) in four states—Iowa, Minnesota, Montana, and North Dakota. In addition to the methodological caveats noted in Chapter 3, some cautions are in order in interpreting the data reported in this chapter:

- Our economic data are a snapshot of carefully selected farms at one point in time. Thus, these data may reflect short-term price fluctuations that influence the relative profitability of conventional and sustainable farms.

- Because our results are based on data at one point in time, they may not represent how our study farms would perform over several years. This is especially true for sustainable farms whose economic strategy is low-risk and long-term. A full evaluation of this strategy would require a long-term study that includes the risk measures identified by the National Research Council report (items 6 and 7, preceding).

- Financial data were gathered on a whole-farm basis. Thus, these data do not permit detailed analysis of individual crop or livestock enterprises or of the fields of surveyed farms.

For these reasons, our analysis cannot "prove" that one farming system is more viable economically than the other. Instead, it is a look at some economic realities, and it flags areas that need further research and policy attention.

Agriculture is very location-specific, and insights can be lost when averages are made across entire regions. Thus, this chapter focuses on eight specific groups of farms (a conventional group and a sustainable group in each of the four states) and compares each to state averages. This allows meaningful comparison of the financial health of widely varying farming systems across the four states. In addition, pooling these eight groups allows us to see trends that are less visible when the two farm types are compared state–by–state.

Comparing Our Findings to State Averages

Comparison to state averages must be done with great care, for two reasons:

1. Our samples *do not represent all farm sizes.* We excluded very small farms (see chapter 3) to guarantee that we were examining serious economic farm units, not hobby farms. From our sampling of these larger farms, we selected for comparison the most polar cases of distinctly conventional and distinctly sustainable operations. Because our survey excluded the smallest farms, our results cannot be compared directly to state averages, which include these small units.

2. Our sustainable/conventional farm data and the state averages *have different sources.* Our data are from farmers' income tax returns, which indicate only reported cash transactions, supplemented by interview data. State averages are from USDA data, which seek to report *all* income. We partially compensated for the differences between the farmers' tax forms and the USDA figures by adjusting the cash-based income figures from tax returns for changes in livestock and crop inventories and for any changes in prepaid expenses.

Therefore, where we describe findings in relation to state averages, please remember that our results are suggestive but not directly comparable.

This chapter first examines the financial position (solvency and net worth) for the eight groups of farms. Then we look at their performance in 1991 with respect to cost, productivity, profitability, and capital expenditures. For reference, much of the data discussed in this chapter is provided in Table 6-1.

INDICATORS OF FINANCIAL POSITION

First, a couple of definitions are in order:

Assets are all the things of monetary value on the farm—land, buildings, machinery, crops in storage, livestock, supplies, cash, and investments.

Debt is the amount owed on the farm. All eight groups of SUST/CONV farmers reported debt greater than their state's per-farm average. Because SUST groups generally combined fewer assets with this greater debt, they generally were weaker financially than their state averages. They also were weaker than CONV groups, except in Minnesota, where they were quite similar. Net worth (assets minus debt) among the SUST groups consistently was less than state averages.

Two important measures of the financial position of a farm are net worth and solvency. Here is a brief look at each, how it is determined, and what our survey disclosed about each measure for CONV and SUST farmers in the four states.

Net Worth

Net worth is simply the total assets of a farm minus its indebtedness. A large net worth indicates economic strength, which enables farmers to survive in years of financial loss. For example, a farmer who lacks land, machinery, livestock, or work time can rent or hire these resources and operate a farm, but one bad year would end such an operation, unless large cash reserves were available. For this reason, tenant farming creates greater instability than does high-net-worth farming.

Average net worth ranged from $253,238 (SUST farmers in Iowa) to $569,125 (SUST farmers in Montana). In three of the four states, CONV farmers had greater net worth. There are several possible reasons for this difference. Sustainable farms may have lower net worth because of poorer economic performance. Or, their lower net worth simply may reflect their smaller average size compared to conventional farms. Another possibility is that farmers with lower net worth have been attracted to sustainable agriculture as a way to farm with less capital.

Solvency

Solvency is necessary for any farm to survive. To be solvent, a farm must be worth more than any money borrowed using the farm as collateral. The less solvent a farm, the greater is its burden of interest and principal payments.

Debt/asset ratio is one way to measure solvency. It indicates the borrowed portion (debt) of total farm value. We derived debt/asset ratio by dividing the debt of each farm by its assets, and then calculating the average for these farms. The resulting ratio (example: 30:100) is expressed as a decimal (0.30).

For individual farms, here is the significance of various debt/asset ratios:

Table 6-1. Financial and performance measures for groups of conventional and sustainable farmers, 1991. (Standard deviation is shown in parentheses.)

Aspect or Measure	Figure	IOWA Conventional	Sustainable	State Average[a]
Number of farms		52	55	102,000
Acres		578	373	328
		(452)	(304)	
Assets		$580,776	$370,477	$568,252
		($605,460)	($342,260)	
Debt		$122,705	$124,207	$100,840
		($134,872)	($194,816	
Net worth		$463,973	$253,238	$467,412
		($549,387)	($238,835)	
Debt/asset ratio	6-1a	0.25	0.29	0.18
		(0.24)	(0.26)	
Debt/net worth ratio	6-1b	0.58	0.63	0.22
		(0.83)	(0.83)	
Gross farm income		$161,902	$130,520	$109,753
		($128,032)	($141,711)	
Net farm income	6-4a	$6,325	$6,584	$22,462
		($36,670)	($32,820)	
Gross farm income/acre	6-3a	$439.29	$483.29	$334.17
		($615.95)	($774.14)	
Net farm income/acre	6-4b	$15.30	$34.49	$68.39
		($101.54)	($170.35)	
Gross farm income/hour of family labor		$63.93 $96.70	$37.41 $38.81	Not available
Net farm income/hour of family labor	6-4c	$5.01 ($13.10)	$2.22 ($10.07)	Not available
Return on farm net worth		-1% (30)	-10% (20)	4%[b]
Operating ratio	6-2a	0.78	0.76	0.70
		(0.27)	(0.22)	
Sales/assets ratio	6-3b	0.45	0.42	0.22
		(0.42)	(0.35)	
Depreciation expense/sales ratio	6-2b	0.07 (0.09)	0.09 (0.11)	0.10
Interest expense ratio	6-2c	0.06 (0.08)	0.06 (0.09)	0.08
Ratio of net farm income from operations	6-4d	0.04 (0.34)	0.07 (0.24)	0.21

Table 6-1. (Continued)

Aspect or Measure	Figure	MINNESOTA Conventional	Sustainable	State Average[a]
Number of farms		19	22	88,000
Acres		865	471	341
		(553)	(406)	
Assets		$418,937	$400,872	$396,956
		($242,300)	($288,241)	
Debt		$159,471	$145,980	$73,138
		($141,998)	($194,586)	
Net worth		$259,465	$254,892	$323,818
		($166,285)	($173,282)	
Debt/asset ratio	6-1a	0.37	0.29	0.19
		(0.24)	(0.25)	
Debt/net worth ratio	6-1b	0.84	0.77	0.23
		(0.78)	(1.33)	
Gross farm income		$187,337	$136,066	$89,413
		($144,786)	($126,003)	
Net farm income	6-4a	$29,675	-$72.00	$21,363
		($40,282)	($36,628)	
Gross farm income/acre	6-3a	$216.80	$291.60	$262.26
		($102.80)	($185.70)	
Net farm income/acre	6-4b	$27.70	$10.70	$62.66
		($50.90)	($65.00)	
Gross farm income/hour of family labor		$52.00	$31.50	Not available
		$36.50	$49.10	
Net farm income/ hour of family labor	6-4c	$5.00	-$1.40	Not available
		($16.30)	($11.10)	
Return on farm net worth		7%	-15%	2%[b]
		(48)	(21)	
Operating ratio	6-2a	0.65	0.84	0.66
		(0.23)	(0.50)	
Sales/assets ratio	6-3b	0.51	0.36	0.22
		(0.46)	(0.23)	
Depreciation expense/sales ratio	6-2b	0.14	0.14	0.11
		(0.19)	(0.13)	
Interest expense ratio	6-2c	0.11	0.08	0.09
		(0.11)	(0.09)	
Ratio of net farm income from operations	6-4d	0.11	-0.06	0.25
		(0.3)	(0.63)	

Table 6-1. (Continued)

Aspect or Measure	Figure	MONTANA Conventional	MONTANA Sustainable	State Average[a]
Number of farms		24	28	24,800
Acres		2,929	5,314	2,431
		(4,839)	(10,234)	
Assets		$533,418\	$703,572	$851,137
		($322,107)	($467,426)	
Debt		$113,984	$136,270	$104,766
		($110,723)	($171,624)	
Net worth		$431,505	$569,125	$746,371
		($277,726)	($460,663)	
Debt/asset ratio	6-1a	0.19	0.24	0.12
		(0.17)	(0.22)	
Debt/net worth ratio	6-1b	0.29	0.48	0.14
		(0.32)	(0.65)	
Gross farm income		$113,296	$155,123	$89,746
		($104,196)	($208,118)	
Net farm income	6-4a	$37,694	$21,916	$20,782
		($54,623)	($46,845)	
Gross farm income/acre	6-3a	$100.69	$53.92	$36.91
		($198.75)	($44.43)	
Net farm income/acre	6-4b	$13.30	$2.04	$8.55
		($21.17)	($26.46)	
Gross farm income/hour of family labor		$26.55	$38.99	Not available
		$20.64	$79.29	
Net farm income/ hour of family labor	6-4c	$8.82	$4.20	Not available
		($11.33)	($11.66)	
Return on farm net worth		5%	-11%	2%[b]
		(24)	(30)	
Operating ratio	6-2a	0.59	0.75	0.75
		(0.33)	(0.34)	
Sales/assets ratio	6-3b	0.29	0.22	0.10
		(0.26)	(0.18)	
Depreciation expense/sales ratio	6-2b	0.10	0.11	0.11
		(0.07)	(0.07)	
Interest expense ratio	6-2c	0.07	0.16	0.12
		(0.06)	(0.15)	
Ratio of net farm income from operations	6-4d	0.24	-0.03	0.27
		(0.36)	(0.43)	

Table 6-1. (Continued)

Aspect or Measure	Figure	NORTH DAKOTA Conventional	Sustainable	State Average[a]
Number of farms		38	40	33,000
Acres		2,425	1,677	1,224
		(2,371)	(1,799)	
Assets		$716,419	$412,850	$676,185
		($805,338)	($342,033)	
Debt		$183,097	$141,679	$107,261
		($274,046)	($137,492)	
Net worth		$532,676	$271,614	$568,927
		($650,328)	($307,999)	
Debt/asset ratio	6-1a	0.33	0.41	0.16
		(0.35)	(0.31)	
Debt/net worth ratio	6-1b	0.50	0.81	0.19
		(0.79)	(1.75)	
Gross farm income		$207,447	$105,463	$102,736
		($193,436)	($104,447)	
Net farm income	6-4a	$41,050	$18,848	$20,003
		($69,515)	($35,382)	
Gross farm income/acre	6-3a	$119.07	$100.43	$83.93
		($87.38)	($113.88)	
Net farm income/acre	6-4b	$26.45	$17.38	$16.34
		($42.71)	($21.07)	
Gross farm income/hour of family labor		$49.46 $39.29	$26.24 $28.68	Not available
Net farm income/ hour of family labor	6-4c	$10.37 ($18.82)	$6.22 ($10.43)	Not available
Return on farm net worth	6-2c	3% (15)	10% (22)	2%[b]
Operating ratio		0.63	0.67	0.72
		(0.30)	(0.31)	
Sales/assets ratio	6-3b	0.41	0.29	0.16
		(0.9)	(0.26)	
Depreciation expense/sales ratio	6-2b	0.09 (0.08)	0.13 (0.11)	0.13
Interest expense ratio	6-2c	0.09 (0.08)	0.13 (0.11)	0.10
Ratio of net farm income from operations	6-4d	0.29 (0.52)	0.24 (0.39)	0.21

[a]State averages are from *Economic Indicators of the Farm Sector: State Financial Summary, 1991.* Agriculture and Rural Economy Division, Economic Research Service. U.S. Department of Agriculture ECIFS 11-2.

[b]Rate of return on current income; does not reflect returns based on capital gains or losses.

- Greater than 1.00: the farm has debt greater than its value; therefore the farm is insolvent.

- Greater than 0.4: the farm may be burdened with debt repayment and may be unable to take advantage of opportunities because of its debt load.

- Between 0.4 and 0.1: this is considered the desirable range for farms.

- Less than 0.1: the farm may not be taking advantage of opportunities.

We found debt/asset ratio ranging from 0.19 (for the Montana CONV group) to 0.41 (for the North Dakota SUST group). These ratios are slightly greater than state averages (Figure 6-1a), but not high enough to cause serious financial stress if the farms are financially healthy otherwise.

Debt/net worth ratio is another way of measuring solvency. We

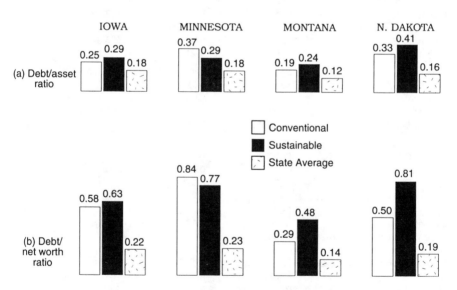

Figure 6-1. Indicators of farm solvency (1991).

derived this by dividing the debt of each farm by its net worth (assets minus debts) for each farm and then calculating the average for these farms. The result is expressed as a decimal.

A value less than 1.0 (for example, 0.57) means that the farmer has a greater investment in the farm than creditors have. A value greater than 1.0 means that creditors have a greater investment in the farm than the farmer does. We found averages from 0.29 to 0.84, so the average owner has more invested than the creditor (Figure 6-1b).

1991 PERFORMANCE OF FARMS IN THE STUDY

Next, we look at the performance of these eight subgroups in 1991 with respect to cost, productivity, profitability, and capital expenditures.

Cost

Operating ratio is the cost per dollar of product. Specifically, it is the total production expenditure for each dollar of gross farm income (production expenditure excludes depreciation and interest). For example, if production expense is 72 cents per dollar of gross income, then the operating ratio is 0.72. If the operating ratio is 1.0, the farmer is spending every penny of gross farm income directly on production.

Operating ratios should be below 0.80 per dollar of expenses to allow money for family living and profit (Farm Financial Ratios, *Doane's Agricultural Report*, July 12, 1992). All the groups we studied are comfortably within this range, except the Minnesota SUST group at 0.84 (Figure 6-2a). Operating ratios for the SUST groups were above state averages in Iowa and Minnesota, equivalent in Montana, and less in North Dakota.

Values for the four CONV groups surveyed are lower than state averages, except in Iowa. This indicates CONV groups were able to control cost and thus be very competitive within their respective states. The Minnesota CONV group (0.65), the Montana CONV group (0.59), and the North Dakota CONV group (0.63) are well-positioned for excellent profitability.

One expectation of sustainable agriculture is reduced dependence on purchased inputs. This means less expense, and therefore implies a lower operating ratio. This is true in Iowa, where SUST farmers had an operating ratio of 0.76, compared to 0.78 for CONV farmers. However, the operating ratios of SUST farmers were greater than those for CONV farmers in the other three states.

These data show that the lower per-acre input purchases expected on sustainable farms do not necessarily translate into lower operating ratios.

Depreciation expense/sales ratio is the cost of depreciation per

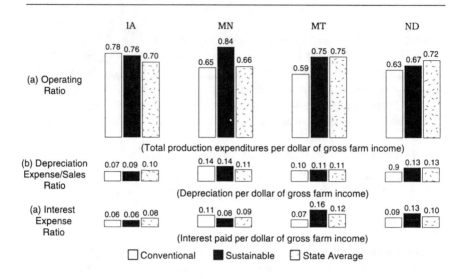

Figure 6-2. Indicators of production cost (1991).

dollar of annual gross sales. It is the portion of gross farm income that must be set aside for replacing worn-out capital items, if the farm is to maintain its current machinery capacity.

The greater the depreciation expense/sales ratio, the more equipment and buildings a farm has per sales dollar. For a farm to be profitable, the depreciation expense/sales ratio must be kept low. Otherwise, a farm could become "capital-poor" or "machinery poor" ("poor" because too much money is tied up in machinery or buildings). Too-low values may indicate too little investment in machinery and buildings to be as profitable as possible; more common is over-investment.

Depreciation expense/sales ratios ranged from 0.07 to 0.14 overall (Figure 6-2b), and the CONV groups fall in this range. For SUST groups, the portion of gross income attributed to depreciation was 0.09 to 0.14. State by state, rates for the SUST group were greater except in Minnesota. This contributes to lower financial performance because a greater percentage of gross income is needed to replace buildings, equipment, and machinery.

What causes the generally greater depreciation expense/sales ratio of SUST farms? There are several possible explanations:

- SUST groups use less commercial fertilizer, so fertility for growing crops must be supplied another way, usually with animal manure and green manure. Animal manure requires application machinery that otherwise would not be needed.

- In the more arid regions of the Great Plains, green manuring, which involves plowing a nitrogen-enrichment crop into the soil every few years, means that land is being farmed without cash return, indirectly raising the depreciation expense.

- Sustainable farms have more crops and enterprises, which requires a greater number of implements.

- Sustainable farms generally have fewer acres in high–income cash–grain crops, which translates to a lower denominator (sales) in the depreciation expense/sales ratio.

Interest expense ratio is the amount of gross farm income paid to lenders as interest on debt. The greater this number, the more difficult it is for the farm to be economically viable. A healthy indicator is a ratio not exceeding 0.10; a reasonable ratio is 0.10 to 0.20 (Farm Financial Ratios, *Doane's Agricultural Report*, July 12, 1992).

For all four SUST groups, an acceptable portion of gross income went toward interest (0.06 to 0.16, Figure 6–2c). Three groups of farmers had an average interest expense ratio above 0.10; Minnesota CONV farmers (0.11), Montana SUST farmers (0.16), and North Dakota SUST farmers (0.13).

Except for Minnesota, all CONV groups reported a lower interest expense ratio than their respective state averages. The Minnesota group reported a reasonably comfortable 0.11, but it was above the state average of 0.09 (Figure 6-2c).

Productivity

Productivity is the *output* of a farm. The best aggregate measure of a farm's output is *gross farm income* (Chapter 4). Other useful measures of productivity show farm output for each unit of input. Such measures include *gross farm income/acre, sales/assets ratio* (output per dollar invested), and *crop yield/acre.*

Gross farm income/acre showed stronger performance for all eight groups than their respective state averages, except the conventional group in Minnesota (Figure 6-3a). Productivity ranged from a gross income/acre of $54 for the SUST group in Montana to $483 for the SUST group in Iowa.

Stronger performance by both groups was expected because we selected only larger farms for our survey. In fact, gross farm income/acre exceeded state averages for CONV groups in all states

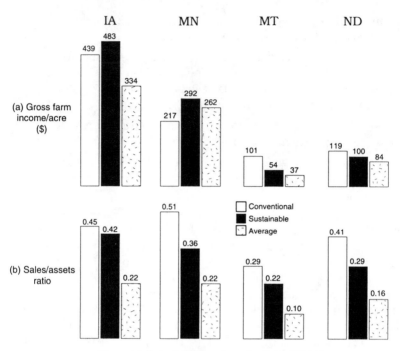

Figure 6-3. Indicators of productivity (1991)

except Minnesota and all four SUST groups. These farmers either have more productive land or farm their land more intensively than is common in their states. More intensive land use could mean producing marketable products such as livestock or high-value crops.

Sales/assets ratio is a measure of the productivity of money invested in a farm. It is the sales yield from dollars invested, expressed as a farm's total sales divided by its total assets. Desirable rates vary by farm type (Fedle and Anderson 1992):

0.25 for cow/calf ranchers ($25 of production for $100 of farm value)

0.45 for cash grain farmers

0.80 for diversified crop/livestock farmers

1.25 for hog farmers—farrow (breeding) to finishing.

Greater sales per dollar of total farm assets generally provide more profit opportunities.

Farms vary regionally, so sales/assets ratios have a broad geographic variation. In our study, Montana had ratios of 0.22 (SUST) and 0.29 (CONV), whereas Iowa reported 0.42 (SUST) and 0.45 (CONV)

(Figure 6–3b). Comparisons among the sales/assets ratios indicate:
- For all four SUST groups, sales/assets ratios were below those for CONV groups in the same state. In North Dakota and Minnesota, SUST groups had much lower sales/asset ratios.
- For all four SUST groups, the sales/assets ratio was above state average.
- For all four CONV groups, sales/assets ratio was at least double the state average, indicating high productivity.

Crop yield/acre measures physical output per acre of cropland. Crop yield/acre varies from region to region, state to state, and crop to crop. Strong economic incentives often discourage livestock production in areas suited to crop production. Therefore, fewer SUST farmers may operate in areas where all the land is suitable for crops.

Of the fourteen comparisons by crop and state, five showed higher yield for sustainable production. Many sustainable farms are located where no single crop has economic dominance, and thus multiple enterprises exist on these farms. Because no single crop is in its optimal environment, yields are expected to be less, unless adjustments are made for land and climate.

For an evaluation of crop yield/acre based on agronomic studies, please see Sidebar 6-2, "In–Field Studies of Sustainable Farm Productivity."

SIDEBAR 6-2

In–Field Studies of Sustainable Farm Productivity

Derrick N. Exner

Productivity of sustainable agriculture is an issue for both farmers and policymakers. To farmers, productivity is an essential component of financial sustainability. To policymakers, productivity is a macroeconomic and political consideration. Many farmers and policymakers believe that more sustainable farming practices bring reduced crop yield and lessened profitability (Bultena et al., 1992). However, little scientific evidence confirms this.

Much of the literature that compares productivity and profitability of conventional and sustainable farming reflects researchers' assumptions and the point-in-time of their investigations. For example, one study showed the U.S. unable to meet international farm-product demand under "organic agriculture"—but their definition employed 1944 agricultural technology, with addition of hybrid seed and crop nutrient inputs at half

---

the level of contemporary U.S. agriculture––a questionable characterization (Olson et al. 1982).

Other studies have used on-farm data. For example, evaluation of two sustainable cropping systems in a five-year South Dakota trial showed corn yields to average less than under conventional systems. The small-grain sustainable system produced lower soybean yields in the first three years, although yield increased as the system approached equilibrium (Smolik and Dobbs 1991). In a Pennsylvania "conversion" experiment, yields from an alternative cropping system equaled or exceeded those of a conventional system after five years (Liebhardt et al. 1989).

But crop yield fails to tell the whole story. In the South Dakota study, despite lesser yields, the alternative systems were more profitable because of lower input cost.

In one of the first comparisons of organic and conventional farms, yields of some Midwestern organic crops were occasionally less, but financial returns of the two systems were virtually identical. This resulted from reduced cost in the organic system. Further, yields in the organic system held up better under stressful conditions (Klepper et al. 1977). A "risk-reduction benefit" also was observed in long-term cropping system trials in Nebraska (Sahs et al. 1988) and South Dakota (Smolik and Dobbs 1991).

Studies in North Dakota, Montana, and Iowa under the Northwest Area Foundation's Sustainable Agriculture Initiative (studies other than the four-state survey reported in this volume) provide data on relative yields of conventional and sustainable farming. The findings also point to other factors that affect the bottom line, going beyond reported crop performance with actual on-farm yield measurements.

North Dakota—Production Under Stress

In North Dakota, researchers compared three farming systems (conventional, organic, and no-till) in three environments (eastern, central, and western North Dakota) on nine farms. Based on two years of data, conventional wheat production had the highest yields and lowest break-even cost in all three regions, although the difference from the organic system was smallest in the western environment, which experiences drought stress (John Gardner, personal communication 1993). The results are shown in Table 1.

These results are consistent with the argument that sustainable farming buffers environmental stress. However, the results are only suggestive, because of the small number of site-years and great weather variability.

Montana—Potential of Green Fallow

Farmers in arid Montana have assumed that planting anything to protect the soil during a noncrop year removes too much soil moisture, stunting growth of the following year's grain crop. Montana on-farm trials tested the potential of low-water-consuming green fallow as a sustainable alternative to bare-ground summer fallow in a very dry climate. These experimental ground covers (legumes) increased rainwater infil-

Table 1. Wheat yields (bushels per acre) and break-even production costs on nine farms in North Dakota (average of 1990 and 1991).

Type	Western		Central		Eastern	
	Yield bu/acre	Break-even cost/bu	Yield bu/acre	Break-even cost/bu	Yield bu/acre	Break-even cost/bu
Convent-ional	26.2	$2.01	56.1	$1.80	66.2	$2.22
Organic	19.6[1]	$2.04[1]	36.2	$2.87	37.2	$2.83
No-till	24.7	$2.83	34.8	$2.17	35.3	$2.67

[1]1990 data only. Crop was destroyed by a hailstorm in 1991.
Source: John Gardner, personal communication 1993.

tration into the soil and contributed nitrogen, as shown in Table 2 (James R. Sims, personal communication 1993).

In most cases, small grains planted after these green manure crops had virtually the same yields as when planted after conventional fallow. Barley exhibited more plump grain and less protein (desirable traits for brewing). Practices based on new technologies like this will increase the options for sustainable farming in drier environments.

Iowa—Importance of Information

In Iowa, researchers found differences in the average corn yields of 93 sustainable and conventional farms, with conventional farms being more productive (Table 3). However, yield differences narrowed when adjusted for two factors that strongly affected yields in 1991—June/July rainfall and the expected corn yield for each farmer's soil type (Table 6-2-2).

The Iowa SUST farmers were drawn from a random sample, plus a sampling of alternative organization members. The alternative organiza-

Table 2. 1991 soil and spring barley characteristics following 1990 green manure crop of Indianhead lentil (Montana farm).

Treatment	Soil		Barley			
	Water Penetration (spring, inches)	Nitrate-N content (spring lb per acre to 4 foot depth)	Yield (bushels per acre)	Test weight (lb per bushel)	Protein (%)	Plump
Green Manure	11.3	53.7	62.5	50.1	11.4	50.6
Bare Fallow	9.9	50.1	60.4	48.8	12.9	39.6

Source: James R. Sims, Personal communication 1993.

Table 3. Corn yields measured in 1991 and yields adjusted for rainfall and soil.

Farm Type and Affiliation	Yield (bushels/acre)	Adjusted Yield[1]	Number of Farms
Conventional	145.1	143.4	44
Sustainable	121.9	128.7	24
PFI & Farm 2000[2]	140.8	134.0	18
OGBA & OCIA[2]	120.0	123.1	7

[1]Corn yield adjusted for rainfall during June-July, 1991, and estimated corn yield for soil type (soil mapping unit) on which field measurements were made.

[2]PFI = Practical Farmers of Iowa; OGBA = Organic Growers and Buyers Association; OCIA = Organic Crop Improvement Association.

tions included two sustainable organizations—*Practical Farmers of Iowa* and *Farm 2000*—and two organic-farming organizations—*Organic Crop Improvement Association* (OCIA) and *Organic Growers and Buyers Association* (OGBA).

Data on four input factors were examined for the 93 Iowa respondents whose corn fields were studied:

- insecticide expenditure
- herbicide expenditure
- grain-drying cost
- cost of applied crop-available nitrogen for the 1989 crop year (nitrogen included purchased nitrogen plus estimated nitrogen available in livestock manure).

SUST farmers averaged $35/acre lower input cost on these four items than CONV farmers. At a 1989 price of $2.20 per bushel of corn, the cost difference in just these four inputs is equivalent to a yield difference of approximately 16 bushels.

Moreover, the reduced purchased inputs were associated with relatively minor yield reductions for those SUST farmers connected to information/support networks. This suggests that these networks helped them farm successfully with fewer purchased inputs. The average corn yields of farmers in PFI and Farm 2000 were greater than those of the unaffiliated farmers and members of the two organic organizations. The organic organizations emphasize marketing support more than crop production methods.

Sustainable Agriculture Can Be as Productive as Conventional

Judging from these studies in three states, sustainable agriculture can be as productive as conventional—*if practitioners have access to information and skills required to successfully substitute their internal resources and management for some purchased inputs.* Successful yields also may depend on the particular cropping system being used, as shown in North Dakota.

These findings argue for agricultural research that is targeted toward more sustainable systems. They also suggest that improved extension education or better organizational outreach might help less-productive farmers gain needed information and management skills.

Profitability

Net farm income is the money generated by a farm to pay for everything beyond production expenses. Net farm income pays for what the farmer provides to the farm: net worth, labor, and management. Since farms have different acreages, labor, assets, and net worth, additional information is gained by looking in greater detail at the profit generated per unit for each of these factors.

To better understand how farmers in our groups performed, we made several calculations: *net farm income/acre, net farm income/hour of family labor,* and *ratio of net farm income from operations.* We discuss each below.

Net farm income provides cash for family living expenses and for building net worth in the farm. We calculated this by adjusting the net farm income reported on farmers' tax returns for changes in the value of crop and livestock inventories and prepaid expenses. (Farmers can prepay expenses to manage their taxable income and to reduce large income tax payments for high-income years. In addition, the volume of crops and livestock in inventory can change, or price changes can greatly alter their value, either raising or lowering a farmer's net worth. To get a true picture of real income in farming in a particular year, such adjustments are necessary.)

The cash flowing into a farm must be used to pay production expenses first. The remaining money--net farm income--then goes to pay for family labor, management skills, and the investment in the farm (net worth).

Net farm income of all SUST groups was below state averages except in Montana, where it was about 5 percent greater (Figure 6-4a). *Net farm income/acre* for three of the four SUST groups was below state averages (Figure 6-4b); all groups were less than half of average, except North Dakota.

This indicates that SUST groups are experiencing much greater production cost/dollar of gross sales. In fact, net farm income/hour of family labor indicates that only in North Dakota are SUST farmers netting more than $5/hour (Figure 6–4c).

This high production cost on sustainable farms, when compared to state averages, reflects in part their greater ratios of debt to net worth. These ratios range from almost three times the state average in Iowa to more than four times the state average in North Dakota (Figure 6-1b). Compared to the average farm in the state, sustainable farms must pay interest on three to four times as much debt per dollar of net worth.

We do not know whether this means farmers with greater debt loads are drawn disproportionately to sustainable farming because of its promise of lower input cost, or whether sustainable farms are somehow accumulating more debt.

In summary, compared to their respective state averages, sustainable farms studied had greater debt and faced more precarious financial circumstances in 1991. But there is considerable variation within the four groups of SUST farmers. Some are doing well (please see Sidebars 6–4, 6–5, and 6–6 at the end of this chapter). On the average, however, their financial performance—in 1991—was insufficient to pay adequately for family labor and capital invested in the operation, while maintaining their net worth.

For all eight groups of farmers studied, net farm income was low relative to total investment, ranging from an average of $41,050 profit for the North Dakota CONV group to a $72 loss for the Minnesota SUST group (Figure 6-4a). In viewing these returns, it is important to remember that a one-year loss is not uncommon in agriculture. Our study reviewed only a single year's data (1991).

Net farm income/acre is the average for each group. It varied from $2 for the SUST group in Montana to $34 for the SUST group in Iowa (Figure 6-4b). However, deeper insight comes when we draw a *per-farm* average, instead of the per-group average.

For example, in the Montana SUST group, if we divide the average net farm income ($21,916) by the average acres per farm (5,314), the per-farm income/acre is about $4. This per-farm average is twice the $2 per-group average, indicating that smaller farms are generating less profit/acre than larger farms. The reason is that, when taking the per-farm average, smaller farms have equal significance to larger farms. This reduces the average because small farms are more numerous.

The Iowa SUST group is just the opposite. When you divide the average net farm income for this group ($6,584) by the average farm acreage (373), the result is about $18 instead of the $34 achieved by calculating income per acre for each farm, and then averaging. In this

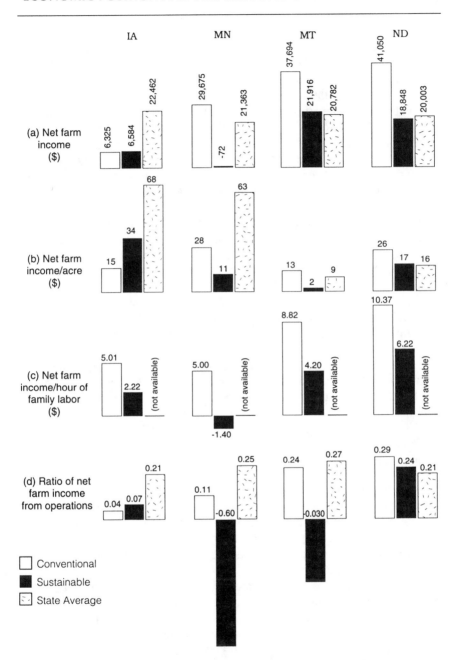

Figure 6-4. Indicators of profitablility (1991).

group, smaller farms are more profitable and pull the average up to almost double the profit per acre.

This insight on the dynamics of farm size also highlights the differences found from state to state in our study.

Net farm income/hour of family labor varied from -$1.40 for the Minnesota SUST group to $10.37 for the North Dakota CONV group (figure 6-4c). This calculation does not include any payment toward what the farmer has invested in the farm, so this is a very low level of income per hour. Three of the groups averaged below $5/hour.

Ratio of net farm income from operations is *net* farm income generated per dollar of *gross* farm income. It indicates the portion of gross farm income that a farmer is able to keep for machinery replacement, family living expense, and return on investment in the farm. This ratio should be positive, of course, and was for all four CONV groups, although averages varied from 0.04 in Iowa to 0.29 in North Dakota (Figure 6-4d).

Among SUST farmers, the ratio of net farm income from operations varied widely. Iowa (0.07) and North Dakota (0.24) reported positive ratios. However, Minnesota (-0.60) and Montana (-0.30) showed negative ratios.

The four groups of CONV farmers reported numbers that varied greatly from their state averages, except Montana (Figure 6-4d). Only the North Dakota CONV group reported a value greater than the state average.

These broad variations are not surprising, considering the small sample size and the broad changes in this variable from year-to-year on individual farms.

Return on Farm Net Worth

Return on farm net worth is the rate of return the farmer receives for the net worth invested in the farm. It is calculated by dividing the return (net farm income minus $5/hour labor) by the farm's net worth. The result is income received from net worth invested, expressed as a percentage.

Of the four groups of CONV farmers, Iowa and Minnesota reported negative returns on the farm net worth. All four of the SUST groups reported negative returns on farm net worth, from –10 to –15 percent. Return on farm net worth was greater than the state average for the CONV groups in Montana and North Dakota, but less in Iowa and Minnesota. These findings suggest that SUST farmers are weaker than their CONV counterparts on this measure of overall economic performance. However, further analysis in Sidebar 6–3 ("Many Factors Influence Farm Performance") indicates that in three of the four states, sustainable practices account for little of the difference between the groups.

SIDEBAR 6-3

Many Factors Influence Farm Performance

Chuck Hassebrook

Farms using sustainable practices had lower returns on net worth in each of the four study states. Why?

One might quickly point to the sustainable practices used. However, we discovered that use of sustainable practices did *not* explain differences in returns among conventional and sustainable farms in three of the four states. Other factors appear to have been responsible for these differences.

We used a statistical technique called *stepwise multiple-regression analysis* to measure the relation between return on net worth of farms and various farm characteristics, including use of sustainable practices, production of livestock and poultry, farm size, and farmers' age and education levels. Regression analysis helps discern complex relations among multiple factors.

In North Dakota, use of sustainable practices explained 15 percent of the difference between farm types in their return on net worth, with sustainable farms displaying the lower returns. Chronological age of the farmers explained another 14 percent of these differences, with older farmers earning the greater returns.

In Montana, we found no relation between sustainable practices and return on net worth. However, we did find that the presence of beef cattle and the amount of gross farm income made significant contributions to explaining such returns. Raising calves was associated with higher returns, and this variable explained 21 percent of the variation in return on net worth. High gross farm incomes also were associated with greater return on net worth, explaining 12 percent of the variation.

A Poor Year for Livestock and Dairy

In Iowa and Minnesota, we found no variable to be consistently associated with return on net worth. Sustainable practices in both states were not associated with low profitability. Other analyses of the data from Iowa and Minnesota suggest that other factors, primarily low 1991 returns on livestock feeding and dairy operations, may explain in part these low returns on net worth experienced by sustainable farmers.

In Iowa, differences in return to net worth between conventional and sustainable farms largely disappeared when the livestock variable was controlled (that is, we compared only sustainable and conventional farms having livestock). Sustainable farms with livestock averaged -11 percent return on net worth, and conventional farms with livestock did

---

about the same, averaging -13 percent. Sustainable farms without livestock fared better, having a -1 percent return on net worth, similar to the 0 percent return obtained by conventional farms without livestock. (It should be noted that only two sustainable farms lacked livestock, a minuscule number.)

Iowa farms with livestock performed relatively poorly in 1991, reflecting lower livestock prices more than deficiencies in their farming practices. That year, the business of finishing feeder pigs (buying young animals and feeding them until they attain slaughter weight) had the second-lowest average profits for any year during the 1984-1992 period. Feeding steer calves yielded the lowest profit for any year since 1985 (Iowa State University Extension, 1993).

It seems paradoxical that calf production (breeding) was quite profitable in 1991—in fact, this explains why cattle were associated with high returns in Montana that year. But the *breeding* of cows to produce calves for sale to feeding operations is a very distinct enterprise from *feeding* calves, which involves buying the animals and feeding them until they attain slaughter weight. The "breeder" farmer benefits from high prices for calves sold for feeding. The "feeder" farmer watches income drop as more is paid for the calves.

The presence of livestock also may have been a factor in the especially poor performance of sustainable farms in Minnesota. Of 22 sustainable farms surveyed, 18 had beef operations (whereas only 3 of 19 conventional farms surveyed had beef operations). Dairy farming undoubtedly was a factor, too; milk prices were lower in 1991 than in any year since 1978 (Wellman 1993). Of these Minnesota farms, 13 of the 22 sustainable farms were dairy farms, versus only 1 of 19 conventional farms.

These findings suggest caution in comparing the lower average returns on the sustainable farms we surveyed to the returns of conventional farms. To the extent that declines in meat and milk prices brought disproportionately low returns to sustainable farmers in 1991, their returns may not reflect long-run performance well. Comparison of whole-farm profitability of sustainable and conventional farms is needed over several years, so that we can separate the effect of temporary price and weather cycles from the economic performance of sustainable practices.

WHERE SUSTAINABLE FARMING STANDS

The SUST groups we studied averaged a quarter to a half million dollars net worth. Even at this level of investment in the farm, the net farm income per hour of labor is less than $5.00/hour, except in North Dakota, where it is $6.22/hour. This is very low compared to many nonfarm jobs, most of which do not require the laborer to make large cash investments to obtain the job and to keep it.

On average, none of the eight groups earned enough to pay for family labor and pay themselves a market rate of return for the net worth they supply to the farm. For the average sustainable farm to survive, family labor must be underpaid, the net worth must be subsidized by other income, or economic performance must improve over 1991 levels. Investment levels of the farms in this study indicate a significant dedication to this way of life.

None of the differences found between the CONV and SUST groups of farmers were consistent across all states. Even within individual states, the tremendous diversity of farms and farming styles generally resulted in much more variation within the CONV and SUST groups than between them.

In general, the financial performance of the four groups of CONV farmers is quite similar to or better than their state averages. SUST farmers generally are not performing as well as their CONV neighbors.

POLICY MATTERS

Despite government intervention in the financial welfare of agriculture, average farming returns have been low. Should the government continue to intervene? If so, what direction should its intervention take?

- One purpose of government intervention is to ensure an excess capacity in agriculture to prevent hunger when shortages inevitably occur. Excess capacity allows for rapid expansion of food production.

- Another purpose is to ensure sustainability. Sustainability assures continued food production for future generations.

Measures to ensure a sustainable level of food production are consistent with the national interest. However, the evidence we present in this chapter indicates that sustainable practices were not economically rewarding during our study year, 1991. Their economic performance that year was not good enough to encourage farmers to use them on a

119

wide scale.

At the same time, there are some bright spots that suggest that sustainable systems can be profitable. In three of the four states we found a group of SUST farmers—the top one-third in terms of their return on net worth—who were earning respectable returns (see Sidebar 6-4, "High-Return Sustainable Farms" at the end of the chapter).

This finding, coupled with plentiful case-study evidence of successful sustainable farms, suggests that sustainable farming may present a greater management challenge (Chapter 9) and that many SUST farmers have not yet learned how to make their systems profitable. (For case-studies of successful sustainable farms, please see Sidebar 6-5, "What Makes Sustainable Farms Successful? Case Studies of Three Minnesota Farms" For a comparative look at financial success, see Sidebar 6-6, "Profitability of Conventional and Sustainable Dryland Grain Farming in Montana.")

If the government wishes to encourage sustainable practices on farms, it must support efforts to improve the economic performance of these systems. This can be done with increased emphasis on sustainable systems in research and education programs. It can be done with policies that increase the cost of environmentally damaging practices, while rewarding use of more sustainable practices. A discussion of the differential impacts of federal farm programs on sustainable and conventional farms, and of policy options that would support sustainable farming, is in Chapter 16.

SIDEBAR 6-4

High-Return Sustainable Farms

Chuck Hassebrook

In our four-state survey, average financial returns for SUST farmers were lower than those for CONV farmers. Yet, some SUST farmers did well in 1991. To find out why, we examined the **return on farm net worth** for the top third of SUST farmers in each state. In three of the four states, these top sustainables earned respectable return on net worth and net farm income:

	Iowa	Minnesota	Montana	North Dakota
Number of Farms	18	7	9	13
Return on Farm Net Worth (percent)	7	0.6	5	11
Net Farm Income (dollars)	29,997	19,376[1]	70,435	49,937

[1]Many of the Minnesota farms are dairy operations, a sector that generally fared poorly in 1991.

In two states, the top third of sustainable farms defied the conventional wisdom that "bigness equals efficiency":

• In Iowa, high-return SUST farmers worked fewer acres and had fewer assets than the middle third. They also had lower *gross* farm income than the low-return third of SUST farmers.
• In Minnesota, the top-return group farmed the fewest acres and had the lowest *gross* income.
• In Minnesota, conversely, the low-return group farmed the most acres and had the highest *gross* farm income.

However, in Montana and North Dakota, high-return sustainable farms were larger on average than the other sustainable farms.

The top SUST farmers used on-farm resources and management practices to reduce purchased inputs. In North Dakota and Minnesota, they scored highest on a scale we used to categorize farmers as *sustainable* in this study.

These top farms practice crop diversity and rotation. In Iowa, the top group scored highest on a scale that measures crop diversification. They planted 46 percent of their crop acreage in crops *other than* the dominant Iowa combination of corn and soybeans. This was more acreage in diversified crops than the 43 percent of the middle group and 39 percent for the low group.

The top group in Iowa also included more crops in their rotations: 4.1 different crops for the high-return group versus 3.7 for the middle-return group and 3.8 for the low-return group. They more often grew oats, committing slightly more acreage than others to this crop.

In Montana, the top-return group raised an average of 4.8 crops, somewhat less than the 5.3 average of the middle group and 5.1 of the low group. In Minnesota, the top group raised more crops (4.0) than the middle (3.4) and low (3.0) groups. No difference was found in the number of crops (3.1) used by the high, middle, and low groups in North Dakota.

Home-Grown Nutrients

The top sustainable group falls on the medium-to-high end of the scale for generating crop nutrients on their farms:

• In Iowa, they purchased 42 percent of the nitrogen used on a typical corn field, versus 34 percent for the middle group and 51 percent for the bottom.

• In Montana, 88 percent of the top group used legumes or green manures in their crop rotation, versus 86 percent for the middle group and 76 percent for the bottom.

• By contrast, the top North Dakota sustainable group used less green manure, averaging just 1 percent of cropland so-fertilized, versus 8 and 9 percent for the middle and low groups. However, the top group used animal manure on a much larger proportion of its acreage—17 percent versus 7 percent for the middle group and 6 percent for the low group.

Fewer Chemicals, But Not Purists

Although top SUST farmers are among the most aggressive in using crop rotations and producing nutrients on their own farms, they are not purists about avoiding chemical use.

In Iowa, the top-return farmers used a greater number of herbicides on corn (1.07 on average) than the middle-return and low-return groups. They also were less likely to forego herbicides altogether, or to limit herbicide use to banding or spot spraying (43 percent of the top group thus limited their use). Yet, the top group kept herbicide cost on corn remarkably low at $5.22/acre. This was somewhat higher than the middle and low groups, but only a fraction of typical herbicide expenditures on corn in Iowa (over $20.00/acre).

Excluding fertilizer, chemical cost averaged a mere $6.52/acre for the top group, slightly below the middle and low groups. However, top-group fertilizer expenditures of $23.22/acre were substantially higher than those for the middle-group ($13.22) and low-group ($17.46).

In Minnesota, the top group spent substantially less on fertilizer and other chemicals per acre than the low-return group. For fertilizer, the top return group spent $13.70/acre compared to $11.58 and $23.54 for the middle and low groups.

In Montana, top-group farmers averaged greater expenditures on fertilizer and other chemicals than the middle and low groups. The Montana results could be influenced by irrigated farms in the sample. If one of the three groups includes a disproportionate number of irrigated farms, it is likely to increase fertilizer and chemical expenditures per acre for that group.

North Dakota is somewhat of an exception. The top group uses herbicides on slightly more acreage (13 percent) than the middle group (11 percent) and low group (9 percent). However, they use fertilizer, insecticide, and fungicide on fewer acres, and they sell more of their products through organic markets. The top group markets 5 percent this way and uses commercial fertilizer on only 9 percent of its acreage, insecticides on only 5 percent, and no fungicides. The top North Dakota group may have enhanced their returns by capturing premiums for organic production.

In summary, although *average* returns in 1991 were weak for sustainable farmers, some farmers earned respectable returns as they aggressively employed sustainable practices.

SIDEBAR 6-5

What Makes Sustainable Farms Successful? Case Studies of Three Minnesota Farms

Jodi Dansingburg

Charlene Chan-Muehlbauer

Douglas Gunnink

Can environmentally sound farming contribute to financial success? For three south-central Minnesota farms studied during the 1992 crop year, the answer is yes. These farms use little or no chemical fertilizer or pesticide, employ other soil-building and environmentally beneficial strategies, minimize capital cost, and emphasize net return more than gross production. The result is an economic return superior to the conventional farms in their area.

We compared these farms to average farms in an area association, the South Central Minnesota Farm Business Management Association (SCMFBMA—See endnote 1). Although gross income for two of the three farms was substantially less than average SCMFBMA farms, *net profit was much greater.* The high net profits were achieved because profit margins (percent of gross farm income retained as profit) were greater than an average of the 90 highest-returning SCMFBMA farms.

Further, these sustainable farms achieved their success on less than half the acreage of the highest-returning 20 percent of SCMFBMA farms. In two of the farms, this also was achieved with fewer cows than average for the region.

Because each farm has unique resources and obstacles, and each farmer has distinct talents, interests, and management ability, there is no single recipe for a successful farming system. Nevertheless, these farms demonstrate the potential for every farm to optimize productivity and bio-

Table 1. Profitability of three Minnesota sustainable farms compared to area averages for the crop year 1992.

Farms	Gross Cash Farm Income[a] ($)	Net Farm Income[b] ($)	Profit Margin[c] (%)	Size of Farm (Acres)	No. of Head[d]
Webster[e]	111,988	63,622	60.4	240	40
Mason[e]	263,508	43,423	21.0	248	75
Elwood[e]	96,683	43,275	48.4	287	28
S. Central Minnesota Average[f]	205,832	28,391	13.3	484	53.5
S. Central Minnesota Highest-Returning 20%[g]	340,641	67,219	18.4	807	53.2

[a]Excludes off-farm income.
[b]Inventory-adjusted profit.
[c](Net Farm Profit)/(Value of Farm Production) x 100
[d] Average number of head on farms with a dairy enterprise.
[e]A pseudonym

[f]Average of 448 farms reported in South Central Minnesota Farm Business Management Association Annual Report.
[g]Average for the highest-returning 20% of the 448 participating farms reported in the South Central Minnesota farm Business Management Association 1992 Annual Report.

logical efficiency, and to improve profits.

Two themes run through these farming operations: *emulating natural systems* and *minimizing production cost*.

Using Natural Systems As a Model

In nature, species interact in systems, like bees and flowers, which work together to manufacture honey. All three farmers work *with* these systems, not against them, to assemble their crops and animals into profitable farms.

Keen Observers. In refining their operations to better work with natural systems, all three operators have sharpened their *observational* skill. Each described how many management decisions have been based on experimentation and observation of soil, weeds, crops, and livestock. Bob Elwood (all names used herein are pseudonyms) described how he decided to feed minerals "free choice" to his dairy herd, rather than force-feeding:

I just kept backing the minerals out of their feed and watched how interested the cows were in eating. The more I backed off on the minerals, the more feed they ate. I can't afford not to take the time to observe these things [soils, crops, weeds, and livestock]. If I farm on a smaller scale [and take the time to observe] I'll make more money.

By experimenting with different soil, weed, crop, and livestock practices and carefully observing the results, each operator has developed integrated farming systems that create healthy environments and profits. Another operator, Kevin Webster, noted:

I don't care as much about NPK levels [levels of nitrogen-phosphorus-potassium chemical fertilizer] as I do about soil microbes and soil health. These organic fertilizers [animal and green manures] slowly release minerals into the soil. On warmer days, the soil microbes are more active, so more minerals become available in the soil. It so happens that on those same warm days, plants are growing faster and require more nutrients. See how Mother Nature fits things together?

All three noted that their soil structure has improved dramatically after substituting for chemicals the judicious use of crop rotation, and managed application of green and animal manures. Their actions have improved the soil's ability to hold water. Now their soils can absorb rains exceeding 2 inches without noticeable runoff, whereas neighboring fields, farmed conventionally, have "streams between every corn row." The third farmer, Jack Mason, observed:

There's something in [my] soil that allows it to hold the water that other people have destroyed by using some of their methods—by some of the [chemicals] that they've put on.

Drought tolerance also appears to be improved by maintaining healthy soils. The farmers commented that their crops appeared to weather recent droughts better than crops of their conventional neighbors.

Humane Treatment of Livestock. The three farms describe here also have modified their livestock production practices to better simulate natural systems. Mason and Webster both are adapting controlled grazing systems to their dairy herds. Webster also raises his pigs on pasture. The animals harvest their own feed, spread their own manure, and stay healthy through exercise and fresh air. "You have to follow the lead of the animals," said Webster. "If you watch them, they will show you what housing and feeding systems they prefer."

Ecological and Agricultural Diversity. A key to using natural systems as a model is *diversity.* Natural ecosystems have diverse soils, plants, and animals that help maintain water and mineral cycles. Each of the three farmers raises varied crops and livestock, creating more natural ecosystems and giving them flexibility against variable market prices:

• Kevin Webster has a 40-cow dairy and a farrow-to-finish hog enterprise. He also markets organic soybeans and small grains.

• Jack Mason has a 73-cow dairy herd (registered holsteins) and raises and sells all his registered bulls and unneeded heifers.

• Bob Elwood has a 28-cow dairy. He sells unneeded heifers and bull calves as beef steers (half the steers are sold directly to consumers, thus increasing profit). He sells some corn and oats, and markets organic soybeans and blue corn.

As Kevin Webster points out, it is the very diversity of his crops and livestock that allows him to piece together a farming system that is productive, yet not overly demanding.

Crop Diversity. Most corn-soybean producers participate in government programs and receive subsidy payments for their corn. This payment is calculated on their *base acres,* which is the area historically planted in corn. Consequently, these corn producers avoid rotating into other crops because they lose base acres, resulting in smaller payments.

All three farms are in a traditional corn-soybean rotation region. Yet, each of these sustainable farmers has made his farming system more profitable by expanding crop rotations, such as growing a mixed crop of field peas and small grain (oats or barley) as part of the rotation. This breaks weed and disease cycles, provides ground cover for erosion control, and does so at very low production cost. The peas add enough protein to the small grain that the harvested crops can be fed to their livestock with little or no supplemental protein.

Webster's livestock gain weight and produce milk so well on this feed that Kevin believes he can raise as much beef, pork, or milk from an acre of barley/peas as he can from an acre of corn, despite corn's greater yield.

One way to encourage more farmers to use beneficial rotations is to change the commodity program to allow planting of beneficial crops without penalizing farmers with loss of base acres (see Chapter 16).

Minimizing Production Cost

Successful limiting of production cost has two components:

1. Minimizing purchased inputs, such as fertilizers and herbicides.
2. Strategically minimizing capital investment.

Minimizing Purchased Inputs. All three farmers have reaped the benefit of emulating natural systems. None purchase commercial fertilizer. Two of them use no herbicide, and the third applies minimal herbicide when mechanical weed control is difficult. In addition to applying manure before planting corn, all three plow in an alfalfa crop for additional fertility.

By relying on farm-generated fertility and pest control, all three farmers saved over $50/acre on fertilizer and chemical cost on their 1992 corn crops, compared to the average farm in their region. (See endnote 2).

Despite the cool, wet growing year in 1992, these three farmers had an average net return/acre on corn of $121, comparing very favorably to the area's average net return/acre on corn of -$7.47 and the average for the top-199 producers of $35. Two of the three farmers achieved their high net returns on corn *without* the assistance of government program payments.

Minimizing production cost carries through to these successful operators' livestock enterprises. All three spend less than the average south-central Minnesota farm on grain and forage for dairy cows, again emphasizing profit over production.

They also feel that restricted chemical use on crops leads to good health among their livestock. Two of the three had below-average veterinary cost for 1992; one spent $31 on veterinary expenses in 1992 for the entire dairy herd. For these operators, the 1992 net return per cow averaged $850, compared to the SCMFBMA average of $387 per cow.

Strategically Minimizing Capital Investment. These farmers also limit investment in livestock facilities:

• Kevin Webster pasture-farrows his sows in the summer (they have their litters in the field, rather than in pens or barns). He uses controlled grazing for both his hogs and dairy animals (see Sidebar 4-2 in Chapter 4, "Controlled Grazing").

• Jack Mason finds it more cost-effective to hire a neighbor to harvest his corn than to invest capital in a combine—plus paying interest on the loan, and maintenance, and repairs.

• Jack also uses controlled grazing to feed his young stock in summer.

• All three farmers have minimized their investment in feed-storage equipment.

• All three run older field equipment to minimize overhead.

• All three have kept their debt levels low by limiting farm size. All the farmers bought their farms and two of the three have paid all of their land debt by using practices described and without the benefit of organic premiums.

By minimizing production cost, these farmers have given themselves resiliency against fluctuating market prices. This resiliency allows them to make a profit under broader conditions than their conventional counterparts.

Conclusion

The three farmers in this study are successful because they manage their farms *as a sustainable whole, not as pieces.* Although each enterprise is environmentally and economically successful, *the overall strength of their farms comes from the synergistic interaction of their soil, crop, livestock, and financial management systems.*

By minimizing production cost, focusing on increasing profits instead of increasing yield, and using natural systems as a model, these three sustainable farms have attained net farm income that is well above the regional average. Their success challenges the conventional wisdom that net farm income can be increased only by increasing gross farm income by means of expanding acreage (requiring capital for land) and livestock confinement.

FOR FURTHER INFORMATION

Land Stewardship Project. 1994. *An agriculture that makes sense: Profitability of four sustainable farms in Minnesota.* A 12-minute video available from Land Stewardship Project, P. O. Box 130, Lewiston, Minnesota 55952, 507/523-3366.

Endnotes

1. Farmers elect to participate in SCMFBMA. Olson and Tvedt (1987) caution that such association farms are not representative of the general farm population. However, SCMFBMA Program Manager Dennis Jackson says that member farms represent the full range of commercial farms in the region, with two exceptions: Association farmers, aged 20 to 60, are somewhat younger than the census average, and SCMFBMA has little or no representation of the small number of very large livestock and crop

farms in the region (personal communication 1993).

2. 1992 per-acre fertilizer and chemical cost for corn: Webster, $0; Mason, $5.42 (for herbicide); Elwood, $12.57 (for soil mineral supplements). Average of SCMFBMA participating farms: $64.86.

SIDEBAR 6-6

Profitability of Conventional and Sustainable Dryland Grain Farming in Montana

Chris J. Neher

John W. Duffield

Does it make economic sense for dryland grain farmers in Montana to use low-input farming techniques in food production? A group of economists from the University of Montana attempted to answer this question in a recently completed study of Montana small-grain production. We interviewed 58 dryland grain farmers, classifying 25 as conventional and 33 as sustainable with a composite index.

This index included three measures:
• The number of sustainable practices used by each farm (mixed cropping, green manures, manure spreading, composting, minimum tillage, windbreaks, subsoiling, soil-sampling, pest monitoring, spot pesticide applications, and biological inputs).
• Per-acre dollar expenditure for chemical fertilizer.
• Per-acre dollar expenditure for chemical pesticides.

Average fertilizer expenditure/acre for the 25 conventional operations was $4.79, compared to $0.98 for the 33 sustainable farms. Similarly, CONV farmers spent considerably more on pesticides ($2.96/acre) than SUST farmers ($1.51/acre).

Conventional and sustainable operations were similar in *average* size (4,308 acres for sustainable versus 3,916 acres for conventional), but they varied widely in *individual* size (280 to 13,200 acres for sustainable, 342 to 12,739 acres for conventional). On average, conventional operations hired 2.60 full-time and 2.64 part-time workers. Sustainable farms made somewhat less intensive use of hired labor, employing an average of 1.91 full-time and 2.09 part-time workers. We compared CONV and SUST groups to determine differences/similarities in profitability, input use, and operational characteristics such as crop diversity and rotation patterns.

We examined profitability differences from three perspectives:

1. *Crop-specific net income/acre* was calculated for both groups. We found very little difference in the per crop-acre net incomes for spring wheat, winter wheat, or barley.

For example, spring wheat, the largest proportion of crop acreage among the sustainable farms (22 percent) and second-largest for conventional farms (16 percent), produced a net income per crop-acre of $56

128

for the sustainables and $53 for the conventionals. Although CONV operators averaged the largest gross income per crop-acre for spring wheat ($91 versus $79 for SUST), they also incurred greater total cost ($38 per crop-acre versus $24 for SUST). The greater cost for CONV operators resulted primarily from their greater expenditure for fertilizer, pesticides, and other unspecified inputs. In general, for the three crops we examined, CONV operators had greater total revenues per crop-acre. However, these greater revenues largely were offset by greater input cost.

2. *Rate of return.* We defined this measure in this manner:

$$\text{Rate of return} = \frac{\text{net farm income (gross farm revenue + government payments - total expenditures)}}{\text{total assets (total value of estate and equipment)}}$$

In this analysis, rate of return measured the profitability on a whole-farm basis, including noncrop activities such as livestock. Although the rate of return (ROR) measure showed sustainable farms to be more profitable (6.1 percent ROR) than conventional (4.0 percent ROR), there was much greater difference within each group than between the groups.

While the average whole-farm rate of return, as discussed above, was greater for SUST farmers, CONV farmers have the larger per-acre return (CONV $36 per crop-acre versus SUST $26 per crop-acre). Sustainable operations compare more favorably using an entire operation assessment (rate of return) than a crop-only basis, because they tend to be more diversified into livestock operations, which adds substantially to overall profitability.

Based on the Montana farms' performance, sustainable operations are competing economically with CONV operations. Our evaluation included only short-term profitability. Long-term benefits of using low-input agricultural techniques are not reflected here. Inclusion of long-term benefits from sustainable measures such as green manure, composting, and legume rotations should strengthen the conclusion that sustainable agriculture is economically viable in Montana.

Table 1. Comparison of profitability measures for conventional and sustainable operations.

Profitability Measure	Crop–Specific Profitability (standard deviation)	
	Conventional	Sustainable
Net income per acre of spring wheat	$52.56 (19.69)	$55.73 (24.75)
	Whole–Farm Profitability	
Rate of return	4.0% (6.4%)	6.1% (8.2%)
Net income per crop–acre	$36.29 (34.38)	$25.93 (27.20)

FOR FURTHER INFORMATION
Duffield, John, Catherine Berg, Douglas Dalenberg, Kay Unger, and Chris Neher. 1993. *Economic evaluation of sustainable agricultural practices in Montana (final report of the Montana Sustainable Agriculture Assessment Project)*. Bioeconomics, Inc., 3699 Larch Camp Road, Missoula, MT 59803, 406/721–2265.

7

Community Trade Patterns of Conventional and Sustainable Farmers

Gary A. Goreham
George A. Youngs, Jr.
Chuck Hassebrook
David L. Watt

Sustainable farmers trade less than conventionals in their nearest communities. They purchase more locally produced farm inputs and livestock, whereas conventionals purchase more crop-related inputs produced elsewhere. More sustainables also trade out-of-state. Increasing the local availability of inputs needed by sustainable farmers and expanding local markets for their products should increase their trade loyalty to nearby communities.

Profound changes in U.S. agriculture are undermining many rural communities. More labor-efficient agricultural technologies have brought rapid farm consolidation, contributing to depopulation of the countryside. Modern transportation has expanded the geographical boundaries of rural trade, facilitating travel to more distant places for goods and services, and in the process, undermining many small towns (Korsching 1985; Labao 1990).

One appeal of sustainable agriculture is that it may help stabilize, or reverse, some adverse economic and social trends in rural communities (Flora 1990; Lockeretz 1986). Central-place theory argues that minimum population thresholds are required for businesses to be profitable, and that such thresholds vary by business type (Borchert and Adams 1963; Goreham et al. 1986). Today, thousands of small towns are declining economically because modern farming practices have pushed the numbers of local farms and farmers below the necessary population thresholds, especially for consumer goods and services.

Agribusinesses often display more staying power in small towns than consumer–oriented businesses because their products (seed, fertilizer, pesticides) are tied to the agricultural land base rather than to population. However, many of these agribusiness firms have consolidated and relocated in recent years as the geographical areas in which rural people trade keep expanding.

A much-cited study of two California communities (Dinuba, with its small family-owned farms, and Arvin, with its large corporate farms and hired labor) suggests that farm type can profoundly affect community viability (Goldschmidt 1978). Although roughly equal in size, the small-farm community of Dinuba had more retail trade and over twice the independent business outlets as the larger-farm community of Arvin.

Based on this and similar studies, some have argued that sustainable agriculture, which encourages smaller farms, can help resuscitate smaller towns. Other studies have challenged this linkage between smaller farms and rural community viability (Barnes and Blevins 1992; Flora and Flora 1988; Skees and Swanson 1988).

For example, Iowa and North Dakota studies found that purchase locations for goods/services were unrelated to farm size or ownership status (Korsching 1985; Goreham et al. 1986). Also, farm size may be only one of several factors contributing to reduced economic viability in large-farm communities (Haynes and Olmstead 1984). Thus, the effects of farm size on rural community viability remain in dispute.

A common argument is that sustainable agriculture benefits local communities in three ways:

1. Sustainable agriculture promotes smaller farms and more people by requiring generally greater management and labor, thereby providing the necessary population thresholds needed by many rural businesses.

2. Sustainable farmers display greater trade loyalty to local businesses, because they are more committed to preserving farm neighborhoods and local communities (Lasley et al. 1993).

3. Sustainable farming retains greater value in the local area by using more inputs produced on local farms or in local communities, rather than contributing to the current loss of value to distant cities and businesses.

Opponents argue that reduced-input sustainable farming is likely to undermine local agribusinesses, especially those supplying fertilizer and farm chemicals. These claims, in turn, typically are countered by arguments that:

- Sustainable farmers will use less of some farm chemicals (of which a very small percentage of profit is retained by local communities), but require other inputs and more services (soil and manure testing, crop scouting, management skills, livestock feeds and services), thus replacing lost "conventional" business opportunities.

- Sustainable farms will require more local consumer goods and services due to larger farm populations.

These arguments suggest that the effects of a more sustainable agriculture could be mixed. Initially, the new farming practices may harm some agribusinesses, but progressive firms will respond by offering new products and services for SUST farmers. (For a community-based assessment of impacts, please see Sidebar 7-1, "Shelby County, Iowa: Sustainable Agriculture and Flexible Agribusiness.") Stores that sell groceries, clothing, and hardware should be bolstered by the larger farm populations expected with sustainable agriculture.

A critical assumption, however, is that sustainable farm families will have sufficient income to contribute to a flourishing local business climate (Dobbs and Cole 1992; Lasley et al. 1993).

SIDEBAR 7-1

Shelby County, Iowa: Sustainable Agriculture and Flexible Agribusiness

Cornelia Butler Flora

Practical Farmers of Iowa (PFI) is a sustainable farming organization working with farmers and agricultural scientists to develop and disseminate sustainable technologies and farming systems. When our study began in 1990, western Iowa's Shelby County had the state's largest concentration of PFI members.

How has this concentration of sustainable farming affected local agriculture practices and agribusiness?

Motivators. PFI members purchase substantially less commercial fertilizer and pesticide than conventional farmers. However, sustainability is not the only motive. Purchased inputs also are declining among many conventional farmers because of increasing cost, declining government price support, and lower corn and soybean prices. These pressures have made farmers more cost-conscious.

According to the U.S. Census of Agriculture, 9 percent fewer farms in 1992 applied commercial fertilizer than in 1987, whereas 27 percent fewer farms applied herbicides. These declines were substantially greater than in a matched county, which had similar agriculture but no farmer member of PFI in 1989.

However, those farms that used chemical inputs applied them to more acres—up to 20 percent for commercial fertilizer and 11 percent for herbicides between 1987 and 1992. These data suggest that it is larger

---

farmers who rely most heavily on purchased chemical inputs. These farmers often buy those inputs outside the county, according to dealers interviewed in Harlan, the Shelby County seat.

Conservation compliance also has motivated many Shelby County farmers to reduce production input. And, a program by Iowa State University Cooperative Extension, in collaboration with the Leopold Center for Sustainable Agriculture and PFI, is showing farmers how to maintain production with fewer purchased inputs. Thus, the Extension agent in Shelby County is convinced that fewer purchased chemical inputs per acre are used.

Pesticide use is difficult to assess. The recent availability of highly concentrated pesticides means that smaller amounts can be equally toxic.

Interest is growing in *integrated pest management (IPM)*. Even conventional farmers no longer feel that a total pest kill is needed to ensure a good crop. *Integrated soil fertility management* also is receiving greater attention. Properly accounting for manure in nutrient management now is common in the county.

Farmers are becoming more careful in applying chemicals, including wearing protective clothing they once dismissed as hot, cumbersome, and unmanly, according to the previous Extension agent. Increasing interest in mixed crop and livestock systems is evidenced by the replacement of that agent when he retired with a cattle–production specialist who has particular expertise in nonconfinement beef operations.

Increased awareness of alternative farming practices has substantially reduced farmers' purchases of production inputs, particularly pesticides and fertilizers. Many now substitute their management effort for capital. This capital formerly was spent locally on commercial products, which supported local businesses, but much of the money left the community to pay the manufacturers and distributors of the products.

Flexible Agribusiness Survives. This large decline in purchases could have spelled disaster for local farm input dealers, but most farmer cooperatives and private agribusinesses in Shelby County are still in business. Less of their gross income now comes from inputs, but their net income has not been seriously affected. Like the local farmers, they have changed their business strategies in the face of the evolving agricultural situation. Like many other American businesses today, flexibility and attention to customers' needs has helped to insure profitability.

One input dealer has gone out of business and another has shifted away completely from pesticide application between 1989 and 1994. (Similar changes occurred in the matched county.) A new agricultural consulting business has been established in town that helps farmers reduce inputs through pest scouting and nitrogen testing, among other services.

Shelby County's agribusinesses offer a different blend of supplies and services today than they did five years ago. As in the past, their choice was to respond to changing demand, or to fail. Unfortunately, the speed of change is greater now, which makes adaptation difficult.

Adding New Services. Agribusinesses increasingly have moved into soil testing by providing a crop specialist who samples soil with a pickup truck-mounted auger and sends the sample to a laboratory for analysis. Another service is scouting field pests, for which agribusinesses hire college students, creating more seasonal jobs.

There is growing competition in nitrogen testing. Kits manufactured in Iowa, now available to farmers by mail from a Colorado distribution point, enable nitrogen testing during the growing season to determine whether additional applications are necessary, and to ensure more judicious use of fertilizer.

According to the local agricultural agent and agribusiness representatives, pesticide application now is done more frequently by custom applicators. Farmers often say that it is better to hire a professional for certain tasks than to hire low-skilled help who may be less conscientious. As a result, agricultural and agribusiness jobs available now more nearly match the skills of Shelby residents, who have a high average educational level. It is noteworthy that custom-applicator jobs often do not require the broad range of skills needed of farm operators, who are running diverse operations.

Diversified Farming, Diversified Agribusiness. Agribusinesses are broadening product lines as farmers become more diversified. After a long decline, livestock production is increasing again in the county. In fact, one Shelby County agribusiness is adding integrated swine and poultry production to keep its feed mills productive. This same firm also has begun supplying computer software and services to farmers. Other businesses have expanded their agricultural credit services.

The important thing is that agribusinesses in Shelby County have not been devastated by declining sales of production inputs, nor by the increase in local farmers who farm more sustainably. Although greater sustainability typically means reduced dependency on purchased inputs, the county's agribusinesses have survived through flexibility. They have shifted product lines and employment patterns to fit the new agricultural production systems being implemented by local farmers.

FOR FURTHER INFORMATION
For further information, see *Sustainable Agriculture, Sustainable Communities: Social Capital in the Great Plains and Corn Belt* (Flora 1994).

We tested some of these arguments. Specifically, we compared the present trading patterns of CONV and SUST farmers to see how they differ, and the importance of these differences for rural community economic viability.

HOW WE MEASURED TRADING PATTERNS

We compared impacts of sustainable and conventional farms on rural communities by examining (a) the type of goods and services purchased (agricultural versus consumer), (b) where these items were purchased, (c) purchase of locally produced goods versus those produced elsewhere, and (d) reliance on out-of-state purchases and sales.

We asked CONV and SUST farmers to identify from a list of 24 items the goods and services they used, which were categorized into farm inputs, livestock inputs, farm machinery, farm services, and consumer goods/services.

We also ascertained the towns where each item usually was obtained, distance in miles to these towns, and whether they were the nearest towns providing the item. This permitted us to compare distances traveled by CONV and SUST farmers to obtain specific goods/services, the size of communities where they traded, and whether these purchases were made in the nearest communities providing the goods/services.

In addition, we asked respondents whether they had purchased any farm inputs directly from other farmers during the preceding 12 months. They also reported goods purchased locally from nonfarmers.

Finally, we asked if they purchased or sold farm inputs or products out-of-state during 1991, and whether these transactions resulted from an unavailability of in-state inputs or markets.

TYPES OF PURCHASES

We examined two broad classes of purchases, those specifically for the farm and those in the general consumer category.

Farm Purchases

We asked CONV and SUST farmers about four types of farm-related purchases:

- Farm inputs (operating loans, seed, fertilizer, and herbicides/insecticides).
- Livestock purchases (livestock, feed, supplies, and veterinary services).
- Machinery purchases (farm machinery, repair, and fuel).
- Farm services (agricultural consulting, crop scouting, custom spraying, and crop insurance).

Conventional agriculture depends heavily on purchase of commercial fertilizers and herbicides/pesticides. In Iowa, all CONV farmers reported purchasing fertilizers and herbicides/insecticides; for SUST farmers this dropped to 83 percent purchasing fertilizer and 64 percent buying pesticides. In North Dakota, nearly all CONV farmers bought farm chemicals, but only half the SUST farmers bought them. No identifiable pattern emerged from the Montana data. (Please see Sidebar 7-2, "Unnecessary Fertilization: A Silent Drain of Dollars from Farm Communities.")

SIDEBAR 7–2

Unnecessary Fertilization: A Silent Drain of Dollars from Farm Communities

Alfred Merrill Blackmer

Recent agronomic studies in Iowa show that most corn producers could increase their profits by applying less commercial fertilizer. The studies utilized new soil and plant tissue tests calibrated to objectively determine the extent to which nutrient availability exceeds crop needs.

Surveys of randomly selected fields have documented the frequency and magnitude of the problem. About half of the cornfields sampled had soil nitrate concentrations greater than twice those needed to attain maximum yields.

Integration of the new soil and tissue tests into on-farm experiments revealed that farmers tend to apply commercial fertilizer in situations where none is needed. Most farmers are surprised to learn that it is not profitable to apply commercially prepared nitrogen fertilizers for corn grown after alfalfa or to soils receiving normal applications of animal manures.

Implementation of the new tests into production agriculture would enable Iowa corn producers to increase their profits while reducing inputs of nitrogen fertilizers by at least $100 million per year. No change in cropping systems would be required. Inputs of phosphorus and potassium fertilizers could be reduced by tens of millions of dollars per year in Iowa alone.

It could be argued that the savings on fertilizer are relatively unimportant, especially if farmers must pay the costs for the soil and tissue tests. However, dollars spent for local consulting services tend to remain and circulate in farm communities. Dollars spent for fertilizer materials tend to flow from farm communities to large corporations outside of farm communities.

The new soil and tissue tests should promote changes in farming systems as farmers gain the ability to determine fertilizer needs in alternative systems on their fields. For example, farmers who learn that manured cornfields require little or no commercial fertilizer may reconsider the merits of integrating crop and animal production.Perhaps the

---

most important information provided by studies with the new tests is that farmers do not know the amounts of fertilizer needed on their fields. Without diagnostic tools like the new soil and tissue tests, farmers have no reasonable means of detecting when unnecessary fertilizers are applied. Unnecessary fertilization cannot be seen or heard by farmers.

Current perceptions concerning fertilizer needs are the product of several decades of cooperation between government and industry to develop new fertilizers and promote their use. These programs have been very effective and have greatly increased the amounts of fertilizer used. The potential beneficial effects of fertilizers on farm productivity and profitability have been widely touted. Relatively little effort has been directed toward evaluating the extent of unnecessary fertilization or assessing its costs to farm communities.

Widespread concern about environmental pollution has generated support for publicly funded programs to improve fertilization practices. These programs have helped farmers recognize the problems caused by inefficient fertilization practices. Much of the effort, however, has been directed toward developing and promoting products that can be sold to farmers. Examples include chemical additives designed to inhibit transformations of the fertilizer in soils. Use of these products often increases the cost of crop production and accelerates the flow of dollars from farm communities.

Many local fertilizer dealers and consultants in Iowa have demonstrated great willingness to offer the new soil and tissue tests as a service to their customers. Recent reports indicate that average rates of nitrogen fertilization in Iowa have decreased by 19% during the past few years. However, most farmers still prefer to pay for extra fertilizer rather than for the tests. The large corporations that manufacture fertilizers have shown little interest in developing or promoting such tests.

The results of the studies point to a clear need for redirecting publicly funded programs toward developing and promoting tools that enable better management decisions on farms. Although it is extremely difficult for farmers and small businesses to develop and promote new technologies that effectively reduce sales of fertilizers and chemicals, they can utilize and advance these new technologies after they are developed and accepted.

The flow of information and technologies made available to farmers clearly has great impact on how dollars flow through farms and farm communities. Continued lack of public funding for programs that target unnecessary fertilization will undoubtedly result in a continuing silent drain of dollars from farm communities.

FOR FURTHER INFORMATION

Blackmer, A. M., T. F. Morris, and G. D. Binford. 1992. Predicting N [nitrogen] fertilizer needs for corn in humid regions: Advances in Iowa. In *Predicting N fertilizer needs in humid regions,* ed. B. R. Bock and K. R. Kelley. Bulletin Y226. National Fertilizer and Environmental Research Center, Tennessee Valley Authority, Muscle Shoals, AL 35660.

In Minnesota, most farmers we surveyed bought these inputs regardless of their farming practices. This probably reflects our survey sampling procedure, which in the case of Minnesota selected some farmers in the middle of the conventional-sustainable continuum, rather than confining selection to the ends, as was done in the other states.

Respondents' purchasing patterns are very different for livestock, livestock-related supplies, and veterinary services. In all four states, well over half of SUST farmers reported such purchases; around half of CONV farmers did so. This, of course, is consistent with the differential percentages in the two categories who have livestock operations.

Consumer Purchases

To examine consumer purchases, we asked about common consumer items (groceries, clothing, hardware, automobiles) and services (auto-mobile repair, physicians, hospitals, personal banking, libraries). As expected, most farmers used many of the listed items, and overall only small differences existed in the purchase patterns of the two farm types.

WHERE DO THEY BUY?

We determined the sizes of local towns from census data, and asked respondents if their points of purchase were the nearest communities for each item, and how many miles they traveled one-way to buy each item.

Nearness to Farm

A substantial majority of both farm types most often purchased goods and services in the nearest towns that had the items. However, SUST farmers were less likely to obtain all of their goods and services in these places. In fact, in three of the four states, more SUST farmers traveled to communities other than their closest towns for fertilizer, livestock feed and supplies, and veterinary services.

Distance to Purchase Site

SUST farmers often drove farther than conventionals to purchase farm-related goods/services, depending upon the state. For example, the distances SUST farmers traveled to make livestock purchases ranged from 26 miles (in Montana) to 50 miles in (North Dakota); the distances

CONV farmers traveled ranged from 16 miles (in Montana) to 47 miles (in Minnesota). In Minnesota, however, they traveled similar or lesser distances than CONV farmers.

The two farm types also differed in their average travel distances to obtain consumer items, but these differences are minuscule. For example, to purchase groceries, SUST farmers averaged 12 miles of travel in Iowa and 29 miles in Montana; CONV farmers averaged 11 miles and 27 miles.

Population of Trade Center

SUST farmers in North Dakota and Montana usually travel to larger communities than CONV farmers to purchase many of their farm-related and consumer goods/services. The reverse generally is true in Iowa.

In North Dakota, livestock are purchased by SUST farmers in places averaging about 22,000 inhabitants, but the comparable population is only 11,500 for CONV farmers. In Montana, the average population of communities where livestock is purchased is about 32,000 for SUST farmers and about 16,000 for CONV. But in Iowa, the average trade center population for livestock purchases is about 171,000 for CONV farmers and 150,000 for SUST farmers. (Note: These results do not take into account respondents' proximity to urban areas.)

Summary and Implications of Purchase Location Findings

SUST farmers purchase farm-related goods and services in their nearest communities less often than CONV farmers; they drive farther and to larger communities to obtain these items. However, we found little difference between the two farm types in the purchase of consumer items, either in distance traveled or in the size of towns where purchases are made. The lack of differences might be more a function of the increasing specialization of rural communities than of farm type, especially for consumer items.

While we found no difference between the two farm types in local purchases of consumer items, the growth of sustainable agriculture may help local areas maintain the population thresholds needed to support consumer businesses. The fact of generally smaller sustainable farms implies more farm families, a greater demand for consumer goods and services, and greater potential for business survival. And if the growth of sustainable farms interrupts the rural population loss, this could have greater long-run significance.

The different farm input needs of CONV and SUST respondents is important to their purchase destinations. Given the generally smaller size of sustainable farms, one might expect them to acquire most farm goods and services locally, but the reverse often is true. SUST farmers tend to drive farther to larger towns than CONV farmers for farm-related purchases. This may reflect the need of SUST farmers for items that are unavailable locally (such as livestock supplies and organic inputs). The exception to this pattern is operating loans, which could be related to the size difference of the average SUST or CONV farms. Historically, farm communities have catered to the crop-related needs of CONV farmers, not to the needs of alternative agriculture.

Also, the population threshold required to support livestock-related businesses is much greater than that for crop-related businesses, thus necessitating livestock farmers to drive farther to often-larger towns for such purchases.

SUST farmers are more likely than CONV farmers to bypass smaller communities and purchase farm items in larger, more distant places. Conceivably, as more farmers adopt sustainable practices, local merchants will become more responsive to their needs and thus draw more trade.

PURCHASES OF LOCALLY PRODUCED ITEMS

Some locally purchased goods needed by SUST farmers also may be locally produced, especially farm inputs (animal feed and bedding, manure for fertilizer, and some types of seed). Selling locally produced goods generally provides greater economic benefit to local communities than the sale of "imported" goods, where often only a small fraction of the profits are retained in the community. By stimulating more trade in locally produced items, sustainable agriculture can benefit local economies.

We explored the relation between farm type and farmers' purchases of goods produced within their local farm communities. This entailed identifying farm inputs purchased directly from producing farmers, as well as purchases of other locally produced inputs. As expected, our data show that both SUST and CONV farmers rely mainly on farm inputs produced outside their local areas. Sizeable numbers buy locally produced inputs, but overall, a higher percentage buy imported goods.

Importantly, SUST farmers more often purchase products produced in farm communities. Figure 7-1 shows that, across the four states, 29 to 54 percent of CONV farmers, and 44 to 71 percent of SUST farmers, make direct purchases of *farmer-produced* inputs. From *nonfarmers*, 13 to 36 percent of CONV farmers and 22 to 54 percent of SUST farmers buy locally produced goods.

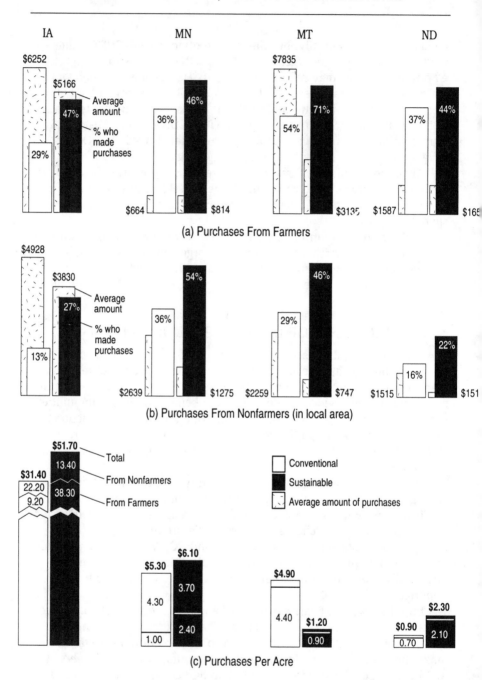

(a) Purchases From Farmers

(b) Purchases From Nonfarmers (in local area)

(c) Purchases Per Acre

Figure 7-1. Local purchases by conventional and sustainable farmers (1991).

Per-Acre Expenditures. SUST farmers more often purchase locally produced items than CONV farmers, but they average smaller expenditures per farm. However, because sustainable farms usually are smaller, their per-acre dollar expenditures on purchases from other farmers tend to be higher than for conventional farms (Figure 7-1).

SUST farmers spend more per acre on inputs from other farmers in three of the four states, though they spend less on other locally produced inputs in three states. Combined, however, the per-acre expenditures for goods and services in these two categories are greater for SUST farmers in three of the four states, ranging from $2.30 to $51.70, versus $0.90 to $31.40 for CONV farmers.

In summary, as expected from their generally larger scale of operations, CONV farmers average greater expenditures than SUST farmers for farmer-produced inputs and other locally produced inputs. However, SUST farmers make these purchases more often and spend more per acre. Expansion of sustainable farming promises to stimulate more sales of locally produced items and thus should help retain more dollars in local economies.

WHO TRADES OUT-OF-STATE?

Sales. Because sustainable and conventional farms use different farming practices and often produce a different product mix, they require different markets. Some SUST farmers who produce nontraditional goods, such as organically grown produce, not infrequently must seek markets outside their local areas.

We asked farmers if they sold products or purchased goods/services out-of-state. We also asked if their out-of-state purchases were prompted by lack of in-state markets or suppliers. The results reported here do not take into account respondents' proximity to state borders.

A majority of both farm types said they buy most of their products in-state. However, some CONV and SUST farmers reported out-of-state sales. In three states, SUST farmers most often reported selling out-of-state (37 to 58 percent), versus 12 to 35 percent of CONV farmers (Figure 7-2).

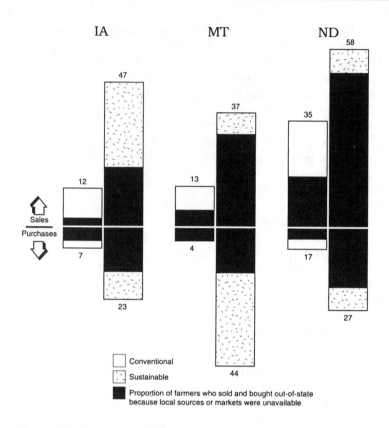

Figure 7-2. Percentage of conventional and sustainable farmers having out-of-state sales and purchases (1991).

In all four states, more SUST farmers, ranging from 33 percent in Minnesota to 80 percent in Montana, reported a lack of in-state markets for their farm products. CONV farmers lacking in-state markets ranged from 20 percent in Iowa and Minnesota to 46 percent in North Dakota.

Items sold out-of-state include organically grown crops and livestock, unique crops (such as black soybeans, blue corn, lentils, buckwheat, and a type of wheat called spelt), and less-common livestock and products (such as goats and goat milk).

Purchases. A large majority of both farm types purchase farm inputs in-state (Figure 7-2). However, in three of the four states, a larger proportion of SUST farmers (23 to 44 percent) than CONV (4 to 17 percent) buy out-of-state.

The unavailability of local or in-state markets for some sustainable farm products clearly affects the potential economic impact of sustainable agriculture on communities. This situation can be expected to diminish over time if the number of SUST farmers increases and more communities provide the products/services they require.

CONCLUSION

We found CONV farmers buying and selling farm-related goods in their local communities more often than SUST farmers. This suggests that sustainable agriculture may be adversely impacting local agribusinesses, at least in the short run. This need not be the case, especially if sustainable farms increase in number and local communities begin more actively catering to their needs. Also important is that SUST farmers most often purchase both farm-produced and locally produced items, thus contributing to value-added activity at the local level.

The generally smaller farm size and larger population base of a sustainable farm economy could help stabilize or increase the number of small–town businesses, employees, and residents. However, if sustainable farms produce lower average net farm income, the potential for local economic growth is jeopardized. In different ways, conventional and sustainable agriculture each support local communities. The nature of this support depends greatly upon which state one considers, and on the items being purchased and sold.

A general composite of purchase patterns of conventional and sustainable farmers is shown in the box on the facing page. Clearly, not all farmers fit this pattern, but it provides a handy comparison.

Purchase Patterns of Conventional and Sustainable Farmers

How do their purchase and sales locations compare?

- The majority of CONV and SUST farmers buy goods and services in local towns, but a larger percentage of CONV farmers do so.
- SUST farmers travel more miles to larger towns to buy goods and services.
- More SUST farmers have out-of-state sales and purchases, because local markets are unavailable.

How do their purchases of locally produced items compare?

- More SUST farmers purchase *other farmers' products* and *locally produced items*.
- CONV farmers tend to spend more per farm on purchases from farmers and from others in the local area; SUST farmers tend to spend more per acre on such purchases.

How do their types of purchases compare?

- More SUST farmers purchase *livestock and livestock-related inputs*.
- More CONV farmers purchase *crop-related inputs*.
- CONV and SUST farmers are equal in their purchase of *consumer goods and service*

What are the implications?

- CONV and SUST farmers both support local communities, but SUST farmers are more likely to make out-of-state sales and purchases.

- If local SUST *farm input* outlets and product markets are developed, SUST farmers' level of support for local towns may increase.

- Assuming SUST farms are smaller, an increase in their numbers may increase the number of farmers, businesses, employees, and town residents. But, if SUST farms have lower net incomes, potential for local economic growth may be constrained or nonexistent.

- If the number of SUST farmers increases, local value–adding economic activity may increase as well.

8

Sustainable Agriculture:
A Better Quality of Life?

Gary A. Goreham
Gordon L. Bultena
Eric O. Hoiberg
George A. Youngs, Jr.
Susan Kathryn Jarnagin
David O'Donnell

This chapter explores differences between conventional and sustainable farmers on several quality-of-life measures--organizational participation, stress, and job satisfaction. Persons in the two farm types have comparable levels of participation in local community organizations. Sustainable farmers tend to be more stressed than their counterparts, especially over financial and farm-management concerns. Both sustainable and conventional farmers favorably evaluate agriculture as an occupation.

INTRODUCTION

A good quality of life long has been held up as an important personal benefit of farming. However, this quality has eroded for many in recent decades with the decline of farm neighborhoods, economic stagnation of rural communities, and intensified financial pressures on farmers. In fact, the often-serious financial and socio-emotional problems of farm families today are well chronicled in the literature (e.g., see Davidson 1990, Rural Sociological Society 1993, and Kramer 1980).

Proponents of sustainable agriculture argue that the adoption of alternative farming practices contributes to a higher quality of life for families. This is attributed to their improved financial returns with sustainable practices, greater job satisfaction associated with the replacement of often routinized work roles with new management challenges, more family involvement in agricultural decisions and activities, and greater participation of family members in social organizations. However, the extent to which the adoption of sustainable practices actually contributes to an improved quality of life remains undocumented (Flora 1990: Kirschenmann 1992; Lasley et al. 1993).

In this study, we tested for differences between the two farm types on three quality–of–life measures––community participation, stress, and job satisfaction.

COMMUNITY PARTICIPATION

The social and economic survival of farm communities hinges on more than attracting farm trade. It also depends on the willingness of residents to participate in and to lead community organizations—farm, religious, civic, youth, and professional.

It has been claimed that SUST farmers are more supportive of their local communities through greater participation in organizations. Lasley et al. (1993) suggest that sustainable agriculture will strengthen farm communities only if SUST farmers are stronger proponents of a "public good" community philosophy, and if their social behavior reflects this. But little objective evidence exists that SUST farmers are more "community-minded" than CONV farmers.

Available time is an important factor in farmer participation in organizations. Chapter 5 showed that SUST farmers and their families average more weekly farm-work hours throughout the year than CONV farm families. Further, SUST farmers and spouses often work as many hours, or more, at off-farm jobs. The longer work weeks for SUST farm families may mean less time available for community activities. However, if SUST farmers are indeed more community-minded, they should find time to participate.

The Iowa and North Dakota surveys included community-participation questions. We asked whether respondents and/or spouses were members of local volunteer organizations, including farm, religious, civic, youth, business or professional, fraternal, service, or school. (We included examples of these organizations to help categorize their responses.) For each group, we also requested their membership duration, proportion of activities attended, and their leadership roles (officer, committee, board, etc.).

Equal proportions of CONV and SUST farmers reported membership in the several organization categories, and their total group memberships are comparable (Figure 8-1). Both types of farmers most often belong to farm organizations, followed by church-related and civic organizations. An average of three to four organizational memberships are reported by each farm type.

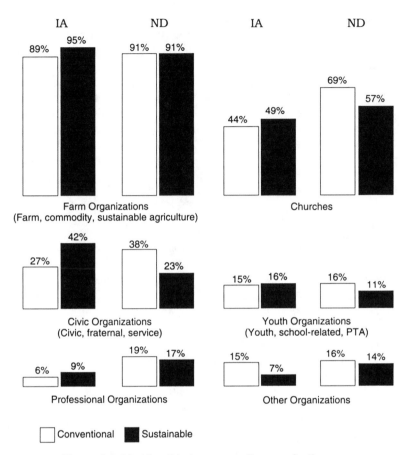

Figure 8-1. Membership in community organizations,
Iowa and North Dakota (1991).

Spouse participation rates (Iowa data only) also show a nearly identical proportion of CONV and SUST people holding organization memberships. Spouses of both farm types most often belonged to church-related organizations, followed by farm organizations.

Few differences between CONV and SUST farmers and spouses were found when examining other dimensions of participation (length of membership, proportion of meetings attended, leadership roles, number of hours spent on organizational activities, and intensity of spouse participation).

The remarkably similar community participation patterns of our respondents does not support the argument that SUST farmers/spouses are more active in their local communities than conventionals. But it is worth noting that they are no *less* active, despite their greater labor on the farm.

Also important in considering community impacts is the number of farm families under conventional and sustainable agriculture that are available to participate in local activities. The industrialization of agriculture has brought a marked depopulation of many rural areas. The likelihood of an increased number of farms, if sustainable agriculture were to become widely practiced (see Chapter 4), suggests that there also would be more persons to participate in community organizations, as well as to shop mainstreets and attend local schools.

FARM-RELATED STRESS

Stress experienced by farm operators is an important quality–of–life indicator. Stress often is induced by an adverse farm economy or weather conditions. For example, many farmers experienced intense stress during the farm crisis of the 1980s, when land prices plummeted, and again during 1993, when heavy rains and flooding complicated field work and caused low crop yields, especially in Iowa and Minnesota.

CONV farmers should be more stressed because (1) they are more vulnerable financially due to greater debt, (2) they use more farm chemicals and thus face greater health risk, and (3) they are less self-sufficient and thus more vulnerable to external control of their farming decisions and financial destinies.

Sustainable farming practices may help reduce some of these stressors. However, sustainable practices could trigger or intensify others. SUST farmers could be stressed by their more complex management requirements, greater labor demands, and tighter scheduling of agricultural activities. Many were attracted to alternative farming because of financial difficulty and a desire to reduce input cost (Chapter 9). Unless their fortunes have brightened appreciably, they face continuing financial stress.

We measured stress among CONV and SUST farmers with 24 items from the *Farm Stress Survey* (Eberhardt and Pooyan 1990) and the *Farming Stress Inventory* (Walker and Walker 1987). Respondents rated each stressor on a five-point scale from "no stress" (one point) to "a great deal of stress" (five points).

The 24 stressors were placed into six general stress categories:

financial, farm management, health, public service/community, environmental, and other. Overall, respondents reported only moderate stress on each item. In fact, most of the average stress scores for each farm type fell slightly below the midpoint of possible scores (less than three points).

The highest stress levels were found for financial items (repayment of farm loans, farm profitability, cost of production, and market prices) and weather conditions. Comparatively high stress levels also were found for some farm-management items, especially compliance with government farm programs, insufficient time to complete farm work, and complexity of farm decisions.

The lowest stress levels generally were reported for adequacy/quality of community services and institutions, such as fire and police protection.

Comparison of stress levels between the two farm types reveals that SUST farmers are more stressed on financial matters, especially concerns about the profitability of their farms, repayment of farm loans, and fears their farms may not be economically competitive. SUST farmers also display more stress about some aspects of farm management, including the timely completion of farm tasks and complexity of decision making.

Also, ironically, SUST farmers reported greater stress on some environmental items, such as groundwater pollution and soil erosion on their farms. This probably reflects the importance of environmental concern in their decisions to adopt sustainable farming, and continuing concern despite their actions to protect environmental quality--see Chapter 9. (CONV farmers likely have similar or worse soil erosion and groundwater contamination problems, but they either are less conscious of them, discount them as a personal threat, or are less attuned to their social cost.)

SUST farmers reported greater stress than conventionals over the uncertain future of their local communities. Again, respondents' stress levels probably say less about their communities' prospects than about differences in the importance that each farm type places on preserving community viability.

JOB SATISFACTION

Job satisfaction is important to the quality-of-life of farm families. We examined (a) how SUST and CONV farmers compare in job satisfaction, and (b) whether SUST farmers' job satisfaction was affected when they adopted alternative farming.

Respondents expressed agreement/disagreement on a five-point scale with several statements about job satisfaction. Their responses suggest several conclusions:

- Nearly four-fifths are "very satisfied" with farming. Importantly, we found no differences in the satisfaction of CONV and SUST farmers; both report high job satisfaction.

- A sizeable majority, nearly three-fifths of both groups, would make the same occupational choice if they were starting over.

- Nearly two-thirds of CONV and SUST farmers in each state have given little or no thought to leaving farming for another job.

- A large majority of both farm types affirmed that farming fulfills their personal goals and contributes to their sense of self-worth.

Comparing Job Satisfaction of Conventional and Sustainable Farmers. Most farm operators, CONV or SUST, are enthused about farming as an occupation. Many feel it provides important personal rewards. Some would not choose this occupation again, but most would. Importantly, there is no evidence that SUST farmers are any less, or more, satisfied with their work than CONV farmers.

The item revealing the least commitment to farming dealt with leaving farming for another job. Whereas a sizeable minority (approximately one-fourth) had entertained such thoughts, other responses indicate that they were spurred largely by financial considerations rather than disenchantment with farm work.

We asked Iowa respondents about the likelihood of a change in job satisfaction as a result of adopting more sustainable practices. Two-thirds of the SUST farmers reported greater job satisfaction now than before adopting the new practices. Only four percent reported becoming less satisfied.

This same favorable outcome is not expected by Iowa's CONV farmers. Three-fifths said their job satisfaction would remain unchanged, and a fourth expected it to decline if they adopted alternative practices.

Sustainable Farmers' Reaction to Adopting Alternative Practices. In unsolicited comments to interviewers or written remarks on their mail questionnaires, some SUST farmers noted a particular attraction of sustainable agriculture: it provides welcome challenges not found in conventional farming. They relished the need and opportunity to better understand biological/ecological processes on their farms, and to draw upon natural principles in producing crops, controlling pests, and pasturing livestock.

Some discussed their previous boredom with "prescription" farming, where they simply applied recommended levels of production inputs with little thought to impact on soil and environment. These farmers drew great personal satisfaction from their on-farm research and insights into farming smarter and more profitably without spoiling the environment.

Some SUST farmers acknowledged the often long hours demanded by their new farming practices and the pressures of completing activities on schedule, but added comments on the importance and value of hard work. They feel that greater labor demands have contributed to their satisfaction with sustainable agriculture and have given them a new sense of self-worth.

Thus, the greater labor demands of sustainable agriculture are not necessarily seen as a drawback by farmers who adopt the new practices. For some, increased labor offsets the need to seek off-farm employment and helps ensure them full employment.

CONCLUSION

Few or no differences are found between CONV and SUST farmers on the three quality-of-life measures. Community participation patterns of CONV and SUST farm operators and spouses are similar. Also, both farm types display a high degree of satisfaction with farming, although the sustainable farmers report improved satisfaction since implementing their new practices. Sustainable farmers have somewhat higher levels of personal stress, especially over financial matters and the complexities of farm management.

9

Adoption of Sustainable Agriculture

Eric O. Hoiberg
Gordon L. Bultena

Sustainable agriculture is not a fixed set of practices/technologies, but involves continuously replacing less sustainable practices/technologies with more sustainable ones. We must understand sustainable farmers' motives for adopting new and often controversial practices, and the barriers that they must overcome, to help both farmers and policymakers make a successful transition to a more sustainable future.

STUDYING TECHNOLOGY AND SUSTAINABILITY

Social scientists long have studied innovative agricultural technologies. Until recently, however, they dwelt more on how technologies were introduced and diffused than on their societal/environmental consequences (Goss 1979; Rogers 1983; Fliegel 1993). Also, past studies have looked at innovations already diffused throughout farming communities, instead of looking forward to inform public policy about the impending impact of new technologies.

A new generation of agricultural research, focused on technologies and practices that have yet to enter the system, or that have been only partially diffused, is now allowing us to predict their social and economic impact.

For commercial innovations, studies show that profit is the primary adoption motive. But how relevant is this finding for *non*commercial adoptions of conservation practices, especially when these adoptions may be prompted more by concern for the "public good" than personal profit? (Pampel and van Es 1977)

Of course, some have questioned whether environmental adoptions are devoid of a profit motive (Nowak 1983). After all, sustainable farming proponents tout these practices as advancing the public good by enhancing environmental and societal values, *while simultaneously improving the profitability* of individual farms.

155

This chapter examines challenges that face farmers as they adopt more sustainable agricultural practices. The topics include (1) their initial awareness and adoption of sustainable practices, (2) motives for adoption, (3) perceived barriers to adoption, (4) primary sources of information that inform adoption decisions, and (5) personal satisfaction/dissatisfaction with the new farming practices.

INITIAL AWARENESS AND ADOPTION OF SUSTAINABLE PRACTICES

Attaining agricultural sustainability requires the adoption of many complex practices, some implemented incrementally over an extended period. Nonetheless, 90 percent or more of the SUST farmers we studied in Iowa, North Dakota, and Minnesota, and 71 percent in Montana, stated that *they had consciously intended to become more sustainable in their farming operations when they first began adopting the new practices.*

Approximately half the SUST farmers in Iowa and North Dakota, and a third in Montana and Minnesota, began using more sustainable practices prior to 1980. Montana farmers are the most recent to adopt, with about half starting after 1985. Smaller percentages in the other states started this late (Iowa 14 percent, Minnesota 33 percent, and North Dakota 21 percent).

These SUST farmers typically were quite young when they introduced sustainable practices into their farming operations. In all four states, nearly half were under age 30, and no more than 8 percent in any state were over age 50. These findings suggest that younger farmers, perhaps less committed to the status quo and more vulnerable to increasing input cost, are the most open to alternative approaches. Younger or beginning farmers are an important target group for efforts to stimulate change in agricultural practices.

MOTIVES FOR ADOPTING SUSTAINABLE PRACTICES

Diverse concerns inspired our respondents to farm more sustainably: the direction of U.S. agriculture, family health, farm profitability, and the environment. Most farmers were influenced by more than one concern. A substantial majority rated economic, environmental, and health concerns as important. Smaller numbers were influenced by personal philosophies or by specific persons or groups, including their families.

To sharpen the picture, we had SUST farmers list their two most important reasons for adopting sustainable practices. Figure 9-1 shows the result by state, in five general categories (economic, health, environment, family/peer group, and philosophy). In Iowa and North Dakota, family health concerns and environmental issues were foremost (although in reverse order). Significantly, in both states, economic considerations ranked third.

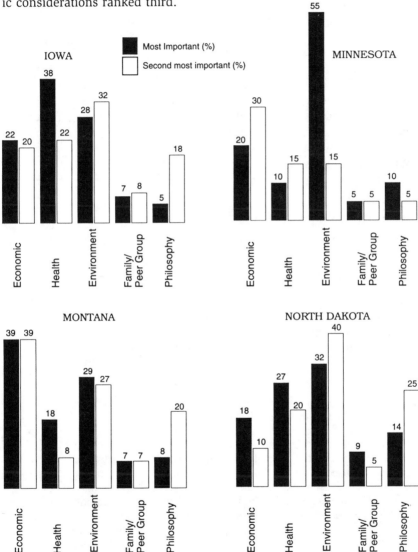

Figure 9-1. Most important reasons for adopting sustainable farming practices (1991).

Minnesota farmers stand out as most highly motivated by environmental concerns, followed by economics. Montana farmers also listed environmental and economic factors as their two most important reasons, but in reverse order.

Philosophical motives ("a new farming system is needed," or "agriculture is moving in the wrong direction") were minor factors motivating most farmers' adoption decisions. Yet, a sizeable minority, especially in North Dakota, included this motive in their top two. (Please see Sidebar 9–1, "Is Sustainability a Philosophical Commitment?")

When making decisions, farmers often seek guidance and affirmation from families or peer groups. The contribution of these groups in shaping adoption decisions is difficult to extract from the multiple motives that affect adoption. For example, family pressure for change may focus largely on health issues, thus explaining why family/peer groups ranked lowest among motivating factors (Figure 9-1).

SIDEBAR 9-1

Is Sustainability a Philosophical Commitment?

Gordon L. Bultena

Eric O. Hoiberg

The alternative agriculture movement does more than promote adoption of new farming practices. It also calls for major reforms in the goals and structure of American agriculture. But how committed to this philosophy of sustainability are farmers who adopt the new practices? Our data from Iowa shed light on this question.

We asked Iowa respondents their opinions on several agricultural issues. Each question contained two opposing statements, one from the conventional standpoint, and the other from the sustainable view. For example:

a. The key to agriculture's future success lies in learning to imitate natural ecosystems and to farm in harmony with nature.

b. The key to agriculture's future success lies in the continued development of advanced technologies that will overcome nature's limits.

Respondents indicated which statement better reflected their personal views. We used 14 pairs of statements, formulated from writings that support conventional and sustainable agriculture (Beus and Dunlap 1990).

The statements dealt with modern agriculture's role in environmental degradation, criteria for measuring success and progress in farming, the relationship of farmers to nature, and their preferred direction for agricultural structure and farm communities. Each statement was given a score from one to five. The lower numbers measured commitment to conventional agriculture and the higher numbers commitment to a more sustainable agriculture. Respondents' cumulative scores spanned the possible

---

range of 14 to 70.

CONV farmers averaged a score of 44, and SUST farmers averaged 60. This indicates major differences in how these two groups view agriculture. Typically, SUST farmers strongly endorsed the philosophical arguments of the sustainable agriculture movement. CONV farmers, however, largely endorsed extant philosophies that long have guided agricultural policy and socioeconomic development in America.

These findings confirm that CONV and SUST farmers are distinguished by much more than different farming practices. They also hold polar views on the desirable direction of future American agriculture and rural life.

PERCEIVED BARRIERS IN ADOPTING SUSTAINABLE PRACTICES

The desire of these farmers to become more sustainable often was tempered by concerns about management, productivity, and profitability. Some questions they commonly asked were:

- *If I reduce my use of synthetic chemical controls, will weeds become more of a problem?*
- *If I further diversify my operation by adding livestock, will my workload expand significantly?*
- *If I reduce the synthetic nitrogen fertilizer on my corn crop, will yield decline substantially?*
- *If I switch to some of these new techniques, will there be reliable information sources to consult?*

These questions reflect concerns that most individuals are likely to raise as they assess the benefits and risks of switching to more sustainable farming.

We gave respondents a list of problems often associated with transition from conventional to sustainable practices and asked them to identify their initial concerns when they began farming more sustainably. Those who were initially "very concerned" about a specific problem were asked whether this concern had increased, diminished, or stayed about the same as they gained experience with sustainable practices. Figure 9-2 summarizes our findings, showing the seven most frequently mentioned initial concerns for SUST farmers.

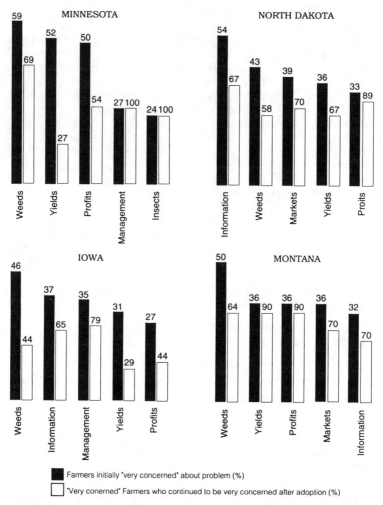

Figure 9-2. Top concerns of sustainable farmers in each state (1991).

- *Weed management* was of greatest initial concern in three of the four states. The proportion of "very concerned" respondents ranged from 43 percent in North Dakota to 59 percent in Minnesota. Once having adopted sustainable farming practices, many of these same farmers remained very concerned about weed management, although their responses varied substantially by state.

- *Yield* was the second-most-common concern of Montanans and Minnesotans, but ranked lower in Iowa and North Dakota. Those "very concerned" about yield ranged from about one-third in Iowa to about half in Minnesota. Importantly, in these two states, only 25 to 30 percent of the farmers who were initially very concerned about yields continued to be so after they acquired some experience, a substantial reduction. However, in Montana and North Dakota, over one-third initially reported high concern about yield, and large proportions of these continued to be concerned (90 and 67 percent).

- *Information availability* was the primary initial concern of North Dakota's SUST farmers and the second concern in Iowa. Over half the North Dakotans and nearly two-fifths of Iowans worried about the availability of alternative-agriculture information when they began. Nearly two-thirds of these remained concerned, even as they became more familiar with alternative agriculture practices. Information was the fifth most common concern of Montana farmers, and was not among the top concerns of Minnesotans.

- *Farm profitability* was the third-most-common concern of Minnesota's farmers, and shared equal weight with yield and markets in Montana. Half the Minnesota SUST farmers expressed a high level of concern about profits, but only about half of these retained this concern. Among Montana farmers, 90 percent remained concerned even after becoming experienced. Profit was the fifth-most-common concern of SUST farmers in Iowa and North Dakota, with a third or fewer respondents listing it initially.

- *Market availability* can be a concern in the transition to sustainable agriculture, especially where specialty crops are important to a diversified cropping system. Slightly over a third of SUST respondents in Montana and North Dakota were initially very concerned about markets, ranking the problem among the top five initial concerns in each state. Market worries did not diminish over time for 70 percent of these farmers. In Iowa and Minnesota, on the other hand, market availability was not ranked among the top five concerns, partly reflecting the lesser crop diversity in these states.

Sustainable farming is characterized by greater diversity, increased biological and cultural control of pests, and more complex nutrient management. Thus, sustainable agriculture often requires greater labor and more sophisticated management skills than conventional agriculture. When asked about the adequacy of their labor supply, a fifth or fewer of SUST farmers reported being concerned initially. In fact, the labor problem did not rank among the top five concerns in any state.

However, larger percentages expressed initial concern about their management skills for handling sustainable practices, especially in Iowa (ranked third) and Minnesota (ranked fourth). Iowans (35 percent) and Minnesotans (27 percent) worried about whether they could successfully operate the new farming systems. In North Dakota and Montana, corresponding figures were 32 and 25 percent (not in the top five). Importantly, these farmers' initial worries about management skill largely remained.

The depth and variety of concerns indicate a bumpy transition into sustainable farming. Some initial concerns diminish over time, as increased experience and new information improve management.

Other concerns, however, are more persistent. Making these the focus of public research and information dissemination should improve the transition of more farmers into sustainability, reducing uncertainty and risk. Also, farm organizations are important in information development and dissemination.

SUSTAINABLE INFORMATION SOURCES

- *How do farmers acquire the knowledge they need to farm in unconventional ways?*
- *Where do they learn the principles of alternative practices?*
- *How is farmers' knowledge influenced by internal and external information sources?*

For decades, most agricultural information has flowed one-way, from scientist to farmer. However, information-sharing among farmers about alternative systems is stimulating development of alternative information sources. This is changing farmers from passive information consumers to *active partners in generating valuable how-to information* (Kloppenburg 1992; Kloppenburg and Hassanein 1993). The generation of new knowledge on farms and distributing it through interacting farmer networks challenges traditional information flow. (Please see Sidebar 9-2, "Sustainable Agriculture: Spreading the Word.")

Sustainable Agriculture: Spreading the Word

Barbara R. Rusmore

John C. Gardner

How do farmers learn about sustainable agriculture? How do they discover management practices that might work in their operations? Both informal farmer-to-farmer networking and sustainable agriculture organizations provide important sources of information.

Informal Farmer-to-Farmer Networks

Farmers and ranchers often learn from each other through conversations over the fence, at the supply store, or at church. This communication is essential to the introduction, adoption, and acceptance of new farming practices.

The informal farmer–to–farmer network was evident in our North Dakota study of farm ecology (Chapter 12). Though the farmers differed considerably in their approach to farming, they listened and studied each other's methods with mutual respect. In Montana, nearly three–quarters of the farmers switching from conventional to sustainable farming turned to fellow farmers for information, a 1987 study showed (Rusmore, 1989).

Clearly, successful promotion of sustainable agriculture must rely on this neighborly exchange.

Sustainable Agriculture Organizations

The more recent Montana survey results reported in Chapter 9 suggest that the most helpful sources have shifted to sustainable farming organizations and farmers' own on-farm research. In addition, the sustainable agricultural organizations involved in the Initiative report recent increases in their memberships. Many farmers in such organizations have been experimenting for years and are a rich source of field-tested innovations. Through formal learning and research networks, sustainable agriculture organizations directly connect new and experienced sustainable farmers.

AERO's Farm Improvement Clubs

Formal farmer-to-farmer networks are particularly important in Montana, where big, open spaces create physical barriers to information exchange and spur feelings of isolation, particularly among farmers trying unique practices.

The Montana farm and ranch improvement club program, started by the Alternative Energy Resources Organization (AERO) in 1989, brings farmers and ranchers together to experiment with sustainable agriculture practices. The program helps farmers and ranchers answer their questions about sustainable agriculture, and builds supportive networks between producers.

Through the program, AERO and the Montana Department of Natural Resources and Conservation award local groups of four or more farmers up to $1,000 for out-of-pocket expenses, like soil tests and seed, on projects of each club's choosing. In addition to the competitive grants, the clubs may request technical assistance from the Soil Conservation Service, Montana State University's extension service and experiment stations, and the Montana Salinity Control Association. Today 22 clubs, representing over 175 farm and ranch families, are finding ways to build soil fertility, lower energy use, manage weeds, improve range forage, expand crop markets and increase crop diversity.

Importantly, each club is farmer-directed, enabling farmers to experiment together with practices in their own operations. "The program was a catalyst," says club member Richard Wrench, who is developing new varieties of garlic and shallots (a type of onion) with his neighbors. "I don't know if I would have done what I did, certainly not in this time frame, if the club hadn't been there."

The work of some clubs has even caused the state experiment station to change its research agenda. In 1991, for example, the Toole County Black Medic Club seeded legume cover crops on summerfallow in mid-summer, rather than early spring, so the annual legume residue would protect the soil from erosion during the winter. Now the agricultural experiment station is testing late-planted legumes as a substitute for summerfallow.

Besides supporting innovative research, the clubs reestablish a feeling of community. Club members get together regularly at club meetings and tours to relieve the sense of isolation that often comes with departing from convention. In exchanging similar ideas and problems, the club members no longer "feel like fruitcakes just because they're trying to conserve resources," says Clint Peck, editor of the *Western Beef Producer*.

By many measures, the farm improvement clubs are a success. According to a recent survey of club members, many farmers and ranchers are implementing more widely the practices they've been experimenting with. The clubs also have given scientists, extension agents and others "a greater acceptance of sustainable agriculture and a better understanding of what farmers are trying to do," says David Wichman, superintendent of the Central Agricultural Research Center.

Clubs are now spread from one end of the state to the other, spanning Montana's entire agroclimatic spectrum, and participants range from conventional grain growers to small organic gardeners to Native American ranchers. "It's really taken off," says Nancy Matheson, AERO agriculture program manager. "Apparently it's something that was needed in Montana."

It's needed in other places as well. New farm improvement clubs started in Idaho, New Mexico, and eastern Washington this year. And a new coalition of private nonprofit organizations, land grant institutions, and state and federal agencies (the Ag Options Network) is working to increase the number of clubs throughout the Northern Intermountain West.

FOR FURTHER INFORMATION
Matheson, Nancy. 1993. AERO Farm Improvement Clubs: A collaborative learning community. *Journal of Pesticide Reform* (Spring), Eugene, Oregon. Also contact: Alternative Energy Resources Organization (AERO), 25 S. Ewing, Suite 214, Helena, MT 59601 (406/443-7272; Fax: 406/442-9120).

We know from prior research that farmers usually rely on multiple information sources when they make decisions. Our findings reveal that farmers find some information sources more useful than others regarding sustainable agriculture.

Rating Information Sources

We asked SUST farmers in the four states to rate the usefulness of information/guidance they received from various sources when they adopted alternative practices (Figure 9-3).

Mass Media (Radio, TV, and Magazines). The media have been important historically in promoting innovative farm products and practices. However, about two-thirds of Iowans and North Dakotans do not rely on radio or TV for information about sustainable agriculture. More Minnesota and Montana respondents tapped these sources, but generally rated them as having limited utility.

Nearly all SUST farmers in Minnesota, Montana, and North Dakota used farm magazines as information sources. A sizeable majority in each state rated magazines as somewhat or very useful. (This question was not asked in Iowa.)

Experts. Experts include local extension agents, area/state extension specialists, university scientists, and Soil Conservation Service (SCS) personnel. Sustainable farmers' reliance on experts differs dramatically among the four states. In Montana, over two-thirds report using each type of expert for information about sustainable agriculture, and a solid majority rates them as somewhat or very useful.

Nearly three-fourths of the Minnesota respondents used local extension agents and SCS personnel, with almost all finding these sources to be somewhat or very useful. Fewer used area/state extension specialists (34 percent) and university scientists (50 percent), but those who did rated them somewhat or very useful.

Sustainable farmers in Iowa and North Dakota rely less on experts. Two-thirds in Iowa had not used extension services for information about sustainable agriculture. A smaller majority had not used scientists or SCS personnel. North Dakotans reported greater expert contact, but about half had not used extension specialists or SCS personnel. Two-fifths had not used extension agents or university scientists.

Agribusiness. In the four states, a fifth to a third of SUST farmers used crop consultants for advice. Dependence on local agribusiness for information varied by state. Approximately two-thirds of Minnesotans and Montanans used them, compared to only 20 percent in Iowa and

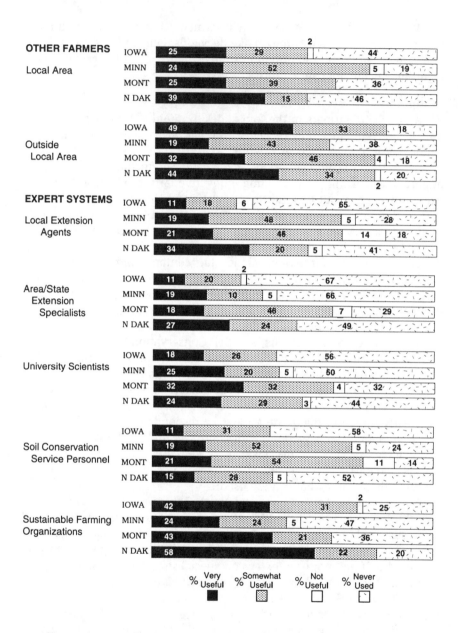

Figure 9-3. Usefulness of information sources on sustainable agriculture (1991).

ADOPTION OF SUSTAINABLE AGRICULTURE

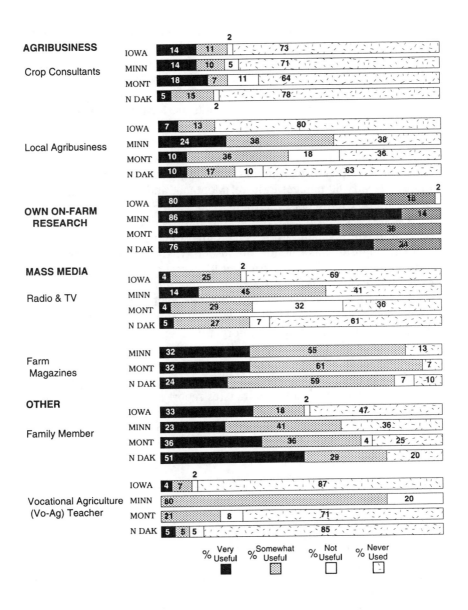

167

37 percent in North Dakota.

Other Farmers, Local and Distant. The importance of information exchange among farmers is well-recognized in diffusing innovative practices. Advice from more experienced farmers in applying a novel idea often is key to successful diffusion of a new practice or product (please see Sidebar 9-2, "Sustainable Agriculture: Spreading the Word"). Whether these support networks are local or regional depends partly on population density.

Except for Minnesota, respondents depended more upon SUST farmers *outside* their local areas than in their own communities. This probably reflects the small number of SUST farmers in any one community. Nearly 80 percent in Iowa, Montana, and North Dakota use nonlocal SUST farmers as information sources. Fewer persons in these three states used local SUST farmers as information sources (from 54 percent in North Dakota to 64 percent in Montana).

The opposite was true for Minnesota, where 62 percent used SUST farmers outside their local areas, but 81 percent used sources in their own communities. *Consistently, in all states, virtually everyone using other farmers for information rated them as somewhat or very useful.*

Sustainable Farming Organizations. Sustainable farming organizations are a formal extension of informal farmer networks. Farmers in all four states use sustainable farming organizations for information and guidance in implementing sustainable farming practices. From a low of 53 percent in Minnesota to a high of 80 percent in North Dakota, SUST farmers in the four states report using the staff, literature, and programs of these organizations. *Almost all of those using these organizations perceive them as somewhat or very useful.* (Please see Sidebar 9-3, "What Role Do Sustainable/Organic Farming Organizations Play?")

SIDEBAR 9-3

What Role Do Sustainable/Organic Farming Organizations Play?

Gordon L. Bultena

Eric O. Hoiberg

What role do sustainable/organic farming organizations play in farmers' adoption of alternative crop and livestock practices? To find out, we compared two groups of Iowa SUST farmers: (a) those holding membership in a sustainable or organic farming organization (*Practical Farmers of Iowa, Farm 2000, Iowa Organic Crop Improvement Association,* and *Iowa*

Organic Growers and Buyers Association), and (b) those not holding membership.

Sustainable/organic organizations generally offer annual meetings, field days, on-farm studies, marketing assistance, social networks, and newsletters. These can stimulate adoption. Also, organization members who have successfully implemented the new practices may be role models for others. Further, sustainable farming organizations often actively promote reform in established agricultural structures and policy.

The adoption motives of the two groups differ. *Nonmembers* most often cite economics (reducing production cost and improving profitability) among their most important reasons for changing practices. *Organization members* are influenced more by noneconomic considerations, including health, environment, and need for fundamental reform in American agriculture.

Organization members rate themselves as more advanced than nonmembers in attaining sustainability in nutrient and weed management. Members also have adopted more sustainable farming practices and have greater desire to learn more about alternative agriculture.

Although SUST farmers often report smaller corn yields than CONV farmers, yield varies substantially. Some SUST farmers equal or even exceed the conventional yield. And members of the two sustainable farming organizations (PFI and Farm 2000), but not the organic groups, have average corn yields that approximate those of CONV farmers, and are considerably greater than the yields of unaffiliated SUST farmers. (Please see Sidebar 6-2)

Iowa's CONV farmers typically endorse philosophies that support the industrialization of American agriculture. Conversely, SUST farmers more often endorse arguments of the alternative agriculture movement. (Please see Sidebar 9-1, "Is Sustainability a Philosophical Commitment?") SUST farmers in organizations generally are more committed than nonmembers to arguments of the alternative movement. Members also display greater concern about their local community's future.

An explanation for our findings may be that farmers holding more extreme orientations and social concerns are more likely to be attracted to membership in sustainable/organic farming organizations. It also is likely that such membership reinforces beliefs and values that diverge from the mainstream.

Conventional and SUST farmers differ in their preferred sources of information. However, information sources of organization members also differ from their unaffiliated SUST counterparts. Nonmembers are more likely to rely on traditional information sources—county extension agents, local agribusinesses, the Soil Conservation Service, and radio/television. Organization members are more attuned to nontraditional information sources—other SUST farmers (especially those outside their local communities), private crop and livestock consultants, and materials from sustainable farming organizations.

Clearly, sustainable/organic organizations are playing an important role in the movement toward a more sustainable agriculture.

On-Farm Research. Sustainable farmers sometimes—accurately—characterize their farms as *living laboratories* where comparative research on alternative cropping systems is a daily fact. There is a strong, widespread commitment to on-farm research across the four-state region. Virtually everyone reported using feedback from their own farming operations as an information source.

Further, *farmers rated their own on-farm research as the most useful in providing information about sustainable farming.* A substantial majority in each state called it very useful (Iowa 80 percent, Minnesota 86 percent, Montana 64 percent, and North Dakota 76 percent).

Why are on-farm research, sustainable farming organizations, and other farmers used more frequently and generally perceived as more useful than experts? This may reflect disinterest in, or bias against, sustainable practices in the agenda of large research and service organizations. It also may be that on-farm research, done in the micro-environment of individual farms, is seen as a superior source of information and feedback (Kloppenburg 1992).

Some feel that the next stage of information generation in sustainable agriculture will flow from the cooperative union that exists among research scientists, technicians, and sustainable agriculture practitioners (National Research Council 1989). However, a fundamental change is necessary, because SUST farmers often are seen as *recipients* of scientific information. Instead, they must be taken more seriously and recognized as full partners in the *generation* of new knowledge about sustainable agriculture (Kloppenburg and Hassanein 1993).

SATISFACTION WITH SUSTAINABLE FARMING

Farmers' satisfaction with their new farming practices seems a good indicator of how successfully these practices are working for them, so we asked respondents how satisfied they were. Those indicating the highest satisfaction levels (very satisfied) ranged from a low of 11 percent in Montana to a high of 44 percent in Iowa.

However, nearly all SUST farmers in the four states were either satisfied or very satisfied with their new practices (84-100 percent). Dissatisfaction was infrequent, ranging from 0 percent in Minnesota to 16 percent in North Dakota. This suggests that neither the concerns of SUST farmers described above, nor their comparatively poorer economic performance, are sufficient to deter them from farming sustainably. However, their economic difficulties and the persistence of some concerns indicates a strong need for greater policy support for their choices.

We next asked respondents to provide reasons for their satisfaction or dissatisfaction. Across the four states, the most common reasons for satisfaction were *reduced soil erosion* and *lower input cost*. In Iowa, the most-mentioned satisfaction factor was *reasonable yield* (33 percent), but yield was less important in other states. *Environmental considerations* also drew mention as a source of satisfaction among Iowa and North Dakota farmers (22 and 14 percent).

Few farmers offered specific reasons for any *dissatisfaction* with sustainable agriculture. Those who did generally cited weed problems, uneven germination, lack of multiple markets, nitrogen deficiency, labor shortages, and the lack of reliable information about sustainable practices.

CONCLUSION

Sustainable agriculture is not a fixed technology. It is an ever-evolving and incremental approach to farming that continuously replaces less sustainable technologies with more sustainable ones (Strange 1990). Thus, SUST farmers in this study might be portrayed more realistically as being "in transition" as they seek to achieve optimum balance among farm productivity, environmental well-being, and long-term profitability.

Our data indicate that farmers who use more sustainable practices made an intentional choice to do so, primarily for environmental, economic, and family health reasons. Further and most importantly, their overall satisfaction with their choice is high. Yet, these farmers seemingly are receiving insufficient support for their efforts from institutional research and extension, agribusiness, and rural development (market infrastructure).

Considering the relatively small number of SUST farmers who benefit from farm programs, and the policy analysis and recommendations that we provide in Chapter 16, current farm policy is failing to provide needed support for larger numbers of farmers to shift to more sustainable practices. It is this critical support that must be provided and maintained to help farmers make the transition of American agriculture into a more sustainable future.

PART

III

LESSONS FROM ALTERNATIVE
APPROACHES TO AGRICULTURAL RESEARCH

Two problems in assessing whole-farm performance are (1) finding sufficiently rigorous and insightful methods to study environmental, economic, and social aspects of agriculture at this scale, and (2) few agricultural institutions or disciplines have a working grasp of what a "farm" is. The methodology gap was bridged with four emerging study techniques: participatory research, case studies and decision cases, the methods of ecological science, and geographical information systems (GIS). Most important, we found the best way to ensure relevant research at the whole-farm scale is to include practicing farmers in the entire research process.

INTRODUCTION

John C. Gardner

Fewer farms dot the countryside today. Fewer farming families practice their profession. Many rural communities are withering into economic and social depression. Yet, our nation enjoys an ample, inexpensive, and high-quality food supply.

The Sustainable Agriculture Initiative's mission was to study this irony and consider the options for fostering an agriculture that is economically, socially, and environmentally successful. Our research focused at the level of farms, farm families, and farming-dependent rural communities, because it is at these levels that the greatest doubt exists about the sustainability of our current agriculture.

To perform the research, the Northwest Area Foundation assembled a team diverse in disciplinary expertise and broad in philosophical orientation. It included academics from the land grant universities, farmers from grassroots farming organizations, and people from non-profit agricultural organizations, both activists and policy analysts.

This team promptly encountered the greatest difficulty of the entire project—*finding rigorous and insightful methods with which to ask the right questions about environmental, economic, and social aspects of agriculture at the whole-farm scale.*

Despite intentions to fully integrate socioeconomic and environmental/agronomic concerns, the team quickly divided along disciplinary lines. Whether academic or farmer, those with an empirical or natural science bent huddled with the agronomists. The social science group attracted the economists, sociologists, and policy analysts. The work proceeded effectively, although this unintended de facto division may have assigned both the topics and the scale that each would address.

The agronomists involved in the Initiative largely dealt with environmental questions: Can yield be maintained while reducing pesticide and fertilizer dependency with locally suitable cultural practices? What is the impact of various farming systems on soil erosion? They weighed alternatives for their ultimate impact on the environment, and often did so at less than farm scale.

The social science group focused on economic and social questions: How does net return compare for alternative practices? What differences in enterprise diversity exist among farms? What are the implications of these findings at the scale of the farm family and farm community?

Another basic difficulty was encountered: *few agricultural institu-*

tions or disciplines have a working grasp of the basic unit of agricultural production we call a farm. If this seems difficult to understand, consider that, from the natural scientist's perspective, an entire "farm" is a burdensome object of study. The precision of natural science requires breaking the farm into smaller, controllable units for study—fields, crops, animals, nutrients, diseases, or insects. These can be compared readily; comparing whole farm units, with their complex and variable systems, is more difficult.

The social science group had a similar problem, but in reverse. Instead of struggling "up" to whole-farm scale, they had to work their way "down" to it. Economic and social issues at a regional, state, or national level had to be related to sustainability at the level of individual farms and their local communities. Socioeconomic methods had to be stretched to accurately profile the economic and social contribution of an individual farm.

We originally sought a scientifically seamless study, one that would assess field-scale farm-management practices for their environmental and economic impact, integrate these findings within the farm, and aggregate them among farms to address community concerns. However, *a methodology gap existed.* Beginning around field scale, and going all the way up to farm-community scale, credible means of study were elusive.

This gap was bridged in part with four emerging techniques, used in projects that paralleled the socioeconomic research described in Part II. To capture and study the essence of the farm, methods adopted included participatory research, case studies and decision cases, the science of ecology, and geographical information systems (GIS). Each made a valuable contribution:

- *Participatory research* was important not so much because it integrated across scale, but because *it brought together farmer and scientist.* It helped assure that science was addressing real problems that farmers face, and it often changed program emphasis. Together, practitioner and theorist studied problems, generally using traditional techniques; but because these studies were directed at on-farm problems, the farmer more easily integrated results at farm scale. (Participatory research is discussed in Chapter 10.)

- *Case studies and decision cases,* techniques borrowed from the business community, are new to the world of the natural sciences and agricultural research. They helped *to capture and integrate the whole of the farm, and to emphasize the process of problem-solving,* rather than its product. (Case studies and decision cases are discussed in Chapter 11.)

- *Ecology* is one of the few natural sciences that has tried to develop integrative concepts and methods. Ecological methods (observational, descriptive) were used *to integrate farm-scale environmental impacts and to identify the costs and benefits of alternative farming practices.* (Ecological research is discussed in Chapter 12.)

- *The Geographic Information System (GIS)* was used to link whole farms into areas that share common resources, such as watersheds. GIS techniques provided a new platform for macro-scale "what if" questions, particularly allocation of scarce resources. (GIS is discussed in Chapter 13.)

No better method exists to assure relevant research at the whole-farm scale than *to include practicing farmers in the entire research process.* Despite the human problems of broad participation in research (individual bias, perception, and personal agenda), the best way to identify problems and create solutions that may have long-lasting benefits for agriculture is to involve the farmers. As the value of this inclusion becomes more widely recognized, so too will the enthusiasm for furthering its use.

The five chapters in Part III describe what we learned through these innovative research techniques. Chapter 14 recounts valuable lessons learned by working together during the Initiative.

10

Use of Participatory
Research in Agriculture

Barbara R. Rusmore
with contributions from
Jodi Dansingburg
Derrick N. Exner
John C. Gardner
Helene Murray

Participatory research includes working farmers. This brings to research the real economic, social, cultural, and management factors integral to successful farming. Five programs in the Initiative demonstrated the value of participatory research in identifying and studying the true needs of farmers and farming communities. Participatory research can be statistically sound, can apply to the whole farm, is useful in addressing a wide range of farmers' daily concerns, encourages innovation and adoption of new ideas, and creates community and leadership opportunities. Participatory research offers the potential for rapidly transforming agriculture.

Within the Northwest Area Foundation's Sustainable Agriculture Initiative, or in association with it, five programs brought farmers and scientists together as partners in research and education. In these programs, the farmers, scientists, and government agency personnel together identified problems for study, determined methodology, collected and analyzed data, and considered the findings and impact on future studies. These programs demonstrate the value of participatory research in identifying and studying the true needs of farmers and farming communities.

This unusual partnership took different forms, depending on who initiated each project—farmers or scientists. But it convincingly showed that joint participation adds a value to studies and can initiate new research directions that are based on farmers' needs.

Participatory research has emerged as a driving force in sustainable agriculture. It is influencing study priorities and spreading adoption of sustainable practices. Although mainstream academic and technical assistance agencies are beginning to recognize the need for sustainable education and research, the ground-breaking work has been done by farmer-driven organizations (Gerber 1991).

The economic crisis of the 1980s catalyzed farmers' organizations to seek alternative ways to farm profitably. They were further guided by environmental concerns and commitment to family farming. (Please see Sidebar 9-3, "What Role Do Sustainable/Organic Farming Organizations Play?") Most farmers experiment regularly, generating significant new knowledge (Jamtgaard 1992b). These farmers' groups organized this natural activity, rallying around common problems and sharing what they learned.

Several participatory research and education groups formed because their sustainability concerns received scant attention from university-based agricultural research and extension (Matheson 1989). These organizations include *Practical Farmers of Iowa (PFI)*, *Alternative Energy Resources Organization (AERO)* in Montana, and the *Land Stewardship Project (LSP)* in Minnesota.

Working with operating farms adds into research the economic, social, cultural, and management factors integral to successful farming (Francis et al. 1990; Rhoades and Booth 1992). Strengthening farmers' research skills and relationships with technical-assistance providers assists farmers in continuously experimenting and improving. *Because participatory research engages this wider agricultural system, it offers the potential for rapidly transforming agriculture.*

THREE MODELS

The three main models of participatory research used in the Initiative's studies originated in work conducted in developing nations. Note that they demonstrate a range of farmer participation:

1. *Participatory Research*—combines cooperative research and adult education to effect social change. Farmers devise the questions, direct the research process, and perform the work to fit their needs. A critical assumption is that, in a just democracy, people *affected* by a problem should be instrumental in *solving* it (Freire 1970; Fals-Borda and Rahman 1991).

2. *Farmer-Back-to-Farmer*—solves production problems and develops new technologies through joint participation of farmers, scientists, extensionists, and the private sector (Rhoades and Booth 1992).

3. *Ethnographic Observation*—studies farms that exemplify a particular farming system in their complete social, economic, and ecological context. Ethnographic observation is performed by an interdisciplinary science team that involves farmers to identify variables and interpret results.

These three models illustrate how differing perspectives can be brought together—those of farmers' organizations (which developed their own research programs, using limited help from scientists and agency personnel), and those of scientists who are seeking ways to apply conventional natural science techniques to test and describe sustainable agriculture.

RESEARCH METHODS USED IN THE INITIATIVE

Each program in the Initiative had a different goal and design, and incorporated elements of the three models. Three of the programs were initiated by farmer-managed organizations prior to the Foundation's Initiative and more closely fit the Participatory Research model. The other two were initiated by university-based researchers in response to the Initiative, and use the other models (Farmer-Back-to-Farmer and Ethnographic Observation).

On-Farm Research and Demonstration—Land Stewardship Project (LSP), Minnesota

Participatory on-farm research was in use with the Land Stewardship Project prior to the Foundation's Initiative. LSP's program is a seasonal and cyclical process of:

1. Winter educational workshops.
2. Farmer-based priority-setting in focus groups.
3. On-farm experimentation, monitoring, and farm tours.
4. Data collection and analysis of results by both farmers and technical staff.
5. Reordering of research needs and resuming experimentation the following season.

Results and observations are published in annual reports available to a wide audience and discussed by farmers during the winter educational workshops.

The Land Stewardship Project's *On-Farm Research and Demonstration Program* seeks to democratize agricultural science by putting environmental and economic investigation in the hands of farmers themselves. More than 75 farm families participated initially in

LSP's on-farm research program. By 1992 it had evolved into the Sustainable Farming Association, with more than 500 farmers organized into six chapters for farmer-to-farmer information-sharing and support. (Sidebar 9-2, "Sustainable Agriculture: Spreading the Word" describes the long-standing communication network among farmers.)

Across the country, participatory research similar to that by PFI, AERO, and LSP is a focus of sustainable agriculture organizations. For more information, two Land Stewardship Project publications are available: *Excellence in Agriculture: Interviews with Ten Minnesota Stewardship Farmers* (Kroese 1988), and *Reshaping the Bottom Line: On-Farm Strategies for a Sustainable Agriculture* (Granatstein 1988).

Farm and Ranch Improvement Clubs—Alternative Energy Resources Organization (AERO), Montana

AERO created a Farm and Ranch Improvement Club program in 1989, to enhance resource sustainability, agricultural profitability, and community development. In this unique, group-centered, participatory research approach, four or more producers form a Farm and Ranch Improvement Club to solve common problems and undertake research projects.

AERO provides the clubs with small grants, organizational and technical assistance, and a supportive learning community. AERO also strongly encourages the clubs to collaborate with staff from their local experiment station, extension, and conservation agency.

This program developed because AERO's network of farmers sought a way to answer their questions and speed up innovation. The program now has 22 clubs, involving about 175 farm and ranch families and other rural community participants. The program's success has led to formal collaboration with supportive federal and state agencies, and was recognized by the 1993 National Distinguished Technology Award.

The club program supports local inquiry and innovation on a wide diversity of topics all across Montana. By holding common meetings and sharing information through farm and ranch tours, publications, and informal contacts, AERO helps match needs with resources and facilitates communication and learning within the network.

One long-time AERO member observed, "we know how [a practice] actually works on our farms, rather than seeing how it works at the experiment station. We have direct hands-on experience with the data so we have more confidence in it...and we are able to share and learn from everyone else too."

Further information appears in Sidebar 9-2, "Sustainable Agriculture: Spreading the Word." More detailed information is available in *Sustainable Agriculture in the Northern Rockies and Plains* (Matheson 1989). Also, please see Sidebar 10-1, "Research by Farm and Ranch Improvement Clubs."

SIDEBAR 10-1

Research by Farm and Ranch Improvement Clubs

Nancy Matheson

On-farm participatory research in Montana's Farm and Ranch Improvement Clubs starts with a group of producers organizing to purposefully investigate a common practical problem. Often they are assisted by a local technical assistance provider in developing and carrying out their research.

What is unusual is that the research questions are established by the producers, and they are involved in all stages of the research project. Small grants (up to $1,000) awarded by AERO have provided the catalyst for many such projects, including:

• Experimenting with alternative legumes
• Ecological response to different grazing strategies
• Feasibility of on-farm vegetable oil processing for an oil/diesel fuel mix
• Green manure versus summer fallow crop rotation trials.

The following examples give some insight into this innovative Farm and Ranch Improvement Club research process:

Lower People's Creek Cooperative

This group, on the Fort Belknap Reservation in north-central Montana, had a problem. They wanted to reduce pesticides used by the Bureau of Indian Affairs (BIA) to control leafy spurge (a weed) on the grassland, and they wanted to try a biocontrol (sheep) instead. Sheep also could be the base for a new economic enterprise, helping to solve another problem, low incomes of co-op members.

During the winter of 1992, the club experimented with controlling spurge on 7,000 acres with intensive grazing by 1,500 sheep. After evaluating their success with the assistance of a Montana State University researcher, the club then increased their herd to 2,300 to control more spurge-infested land, and expanded the enterprise by selling lambs.

The improvement of range and sale of lambs brought this co-op's members closer to their goals. An unexpected benefit was that others, including the BIA, recognized the cost–effectiveness of sheep biocontrol. In fact, BIA now is spending some of its herbicide spraying budget on fencing for tribal members outside the co-op who want to control weeds with sheep. As a result, BIA is spraying fewer acres.

Toole County Club

This club formed in 1990 on the Canadian border of Montana to see if they could incorporate a variety of black medic (a legume) into their grain crop rotation. The research question was "could medic be used to control saline soils and wind erosion, while protecting and enhancing grain crops?"

Each farmer established a test field--which is still being used four years later for experimentation. But the research question has changed. As farmers experimented and compared results from several farms, they realized that the cold, dry climate required alternative approaches. Thus they have worked with the experiment station and others to find new information and understand their research results. Each year the experiments have changed to try different seeds, planting schedules and techniques, and rotations. The question has become: Can we find a legume/small grain rotation that will work in our situation?

Gallatin Valley Growers

In south-central Montana, north of Yellowstone National Park, the Gallatin Valley Growers brought together 15 small, diversified producers of vegetables, fruits, herbs, and flowers. Their challenge was to develop successful marketing of their locally grown produce--in an area that typically gets all its produce from far away. They asked six community business people and university marketing specialists to speak with the group. Then they designed marketing trials, including advertising, networking, and a farmer's market.

Within the first year, Gallatin Valley Growers doubled their sales, and figure they added $150,000 to the area's economy. They also learned that new marketing takes innovation and perseverance--and incorporated a number a changes that will increase their success next year.

Shields Valley Conservation Club

Farmers in this Park County, Montana, club sought alternative cropping systems that are environmentally sound and cost–effective. By talking over options with their soil conservationist, a crop and soils specialist at the university, and other Farm and Ranch Improvement Clubs, this club elected to do research on diversifying grain crops with a marketable pea. They found a customer to buy most of the crop, and discovered that the peas could be marketed fresh to local grocers as well.

However, not all went according to plan. The peas were harder to seed than anticipated, so production was smaller than intended. The buyer so liked the product that he found a more constant supply than the club could offer in this experimental year. The club made changes in production and marketing, making the following year more successful.

In reporting on their experiments, the club leader commented, "As we watched the group trying these projects, we saw that it takes effort and patience, and that these two equal learning. If we take what we learned and apply it next year, we will make great strides in Shields Valley Agriculture."

AERO's Farm and Ranch Improvement Clubs support tremendous diversity in people, geography, goals, and activities. Some clubs pursue on-farm research and demonstration; others conduct market research. But

> the key to success is participation of the farmers in defining the problem, experimenting, discussing the outcome, and making better informed choices about the next step.
>
> **FOR FURTHER INFORMATION**
>
> Further information appears in Sidebar 9-2, "Sustainable Agriculture: Spreading the Word." More detailed information is available in *Sustainable Agriculture in the Northern Rockies and Plains* (Matheson 1989), published by AERO.

Replicated Field Trials—Practical Farmers of Iowa (PFI)

Practical Farmers of Iowa (PFI) is a farmer organization carrying out farmer-initiated and farmer-managed replicated field trials. PFI has collaborated with Iowa State University since 1988 with the goals of (a) promoting farmer-to-farmer information-sharing about profitable, practical, and environmentally sound farming methods, and (b) helping ISU with research in these areas.

A combination of participatory and farmer-back-to-farmer research, the PFI-ISU collaboration began with a joint proposal to the Iowa Department of Agriculture and Land Stewardship for a statewide extension coordinator and farmer-managed, on-farm trials and field days.

From 1987 to 1993, PFI cooperators carried out 348 replicated trials, and total cumulative attendance at PFI field days reached 9,100. What sets these replicated field trials apart from simple demonstrations is that each practice being tested is repeated at least six times in the field. Statistical analysis of the repetitions provides a powerful statement about the reliability of test trials. Dozens of articles in the agricultural press have recognized the PFI-ISU collaborative program, and it received a National Environmental Achievement Award in 1991.

For the Initiative, this existing network of farmer/scientists was used for field-scale sites to test the validity of current nitrogen fertilizer recommendations and soil tests. Along with nitrogen, data on other characteristics such as corn yield and pesticide use were gathered across Iowa, providing a unique and statistically sound database useful to farmers, scientists, and policymakers.

Further information is available in *On-Farm Research: Using the Skills of Farmers and Scientists—Practical Farmers of Iowa and Iowa State University* (Exner 1990).

Farmer/Scientist Focus Sessions—Oregon State University

Farmer/scientist focus sessions (FSFS) are facilitated meetings in which farmers and scientists address selected issues or problems.

Focus sessions take advantage of the creativity and exchange that occurs when farmers and scientists listen carefully and learn from each other.

Using this modification of the farmer-back-to-farmer model, Oregon State University developed several innovative and practical solutions to field and farm-scale problems. Situations where OSU used focus sessions included:

- Investigating complex yet urgent issues that have an incomplete study base (Green-McGrath et al. 1993). Example: food safety perceptions.

- Designing experiments that evaluate treatments on both plot-scale and field-scale. Example: alternatives to Dinoseb herbicide for controlling weeds in snap beans.

- Investigating complex problems of cropping or livestock systems that require interdisciplinary work. Example: disposal of large volumes of cull onions.

Further information is available in *Farmer/Scientist Focus Sessions: A How-To Guide* (Green–McGrath et al. 1993).

Farming Systems Comparison—North Dakota

This study assessed the agronomic, economic, and ecological impact of the most promising models of agricultural production practiced in North Dakota. Three production systems were studied:

1. Conventional (using pesticide, fertilizer, and tillage, with rotations driven mostly by markets and farm program benefits).

2. Conservation tillage (no-tillage with pesticide).

3. Farms tending toward organic production (little or no pesticide and fertilizer, relying on tillage and rotation for pest control and soil fertility) (Clancy et al. 1993).

Scientists assessed these alternatives largely by ethnographic observation, closely examining real farms and farmers who were proven experts in each production system. Nine farms were selected for detailed comparative tests over two years, with scientists visiting the farms bimonthly. The farmers and scientists met as a group at annual project reviews to discuss results, interpretation, and the impact of different management decisions.

Use of these methods kept data collection and interpretation at a delicate balance between the needs of the scientists, who wanted to gather data on specific fields for agronomic comparisons, and the farmers, who saw field-scale decisions as only part of an overall farm-scale management plan.

Further information is available in *Farming Practices for a Sustainable Agriculture in North Dakota* (Clancy et al. 1993).

OUTCOMES FROM PARTICIPATORY RESEARCH

Our studies used participatory research to better identify, study, and assess the problems of farms and farm communities in achieving a sustainable agriculture. We discovered four key features of farmer involvement that are particularly useful for expanding current research and education:

1. *Participatory research can be statistically sound and applicable to the whole farm.*

2. *Participatory research is useful in addressing a wide range of concerns that face farmers daily.*

3. *Participatory research encourages innovation and adoption of new ideas.*

4. *Participatory research creates community and leadership development opportunities.*

Here is a brief look at each feature...

1. **Participatory research can be statistically sound and applicable to the whole farm. Although driven by the concerns of *applied* research, these methods also can make valuable contributions to *basic* research.**

Through collaboration, farmers and scientists learn new information about farming practices and technologies that can be statistically valid and generalizable. The technique is one of applied research *in the context of whole-farm research*—not trials on experiment stations, isolated from real-world decisions, but trials on-farm to see how the practices fit into the farming system.

For example, Practical Farmers of Iowa cooperators, in 71 replicated trials on-farm, successfully reduced their nitrogen rates for corn by 40%, saving an average of $6.68/acre. Ridge-tillage trials substituted mechanical and cultural controls for herbicides, saving $5.60/acre in corn production cost and $5.78/acre in soybeans. Many examples mentioned earlier emphasize whole-farm studies, such as AERO's Farm and Ranch Improvement Clubs and the North Dakota comparison of farming systems.

Scientists in each program reported that *working with producers led research in new directions for sustainable agriculture and generated new study ideas for experiment stations.* This is very important, because basic research has come to be expected only at increasingly smaller scales, such as the current emphasis on molecular biology. By establishing a partnership among farmers, scientists, and extension, study ideas became evident at *all* scales, which is sure to enrich the research agenda.

For example, PFI's work in 49 replicated trials using ridge-tillage with and without pesticides showed that weeds can be controlled effectively and more cheaply without pesticides. This result raises questions about weed ecology that can be explained only through basic research. (An example of such a question is "What environmental factors cause weeds to germinate, and how can we manipulate this to our benefit?)

Rick Exner, of Iowa State University Extension, comments: The research process need not be a "top-down" one. Researchers and farmers can be involved in both basic and applied research. Contrary to conventional thinking, highly creative basic, or "pure" research hypotheses can grow from observations on working operations. In research that hinges on management, working farms are sometimes the only feasible sites.

2. **Participatory research is useful in addressing a wide range of concerns that face farmers daily, from production and resource problems to marketing and community development.**

The Initiative undertook studies at the whole-farm level, which would have been impossible without farmer collaboration. For example, farmers in Oregon and North Dakota kept reminding researchers that field-scale management was strongly influenced by the rest of the farm's need for labor, machinery, and income. This perspective required scientists to consider plant and soil science, economics, and sociology.

Farm management obviously is critical to success and can be studied best with participating farmers. This was exemplified by PFI's collaboration with ISU on narrow-strip inter-cropping, which revealed the potential biological benefits of the practice (improved yield and conservation) more clearly than had plots at experiment stations.

Beyond individual farms, farm community participation also was made possible through participatory research. Nearly one-third of AERO's Farm and Ranch Improvement Clubs address community education and development by providing workshops, forming a marketing cooperative, or starting a farmer's market, for example.

3. **Participatory research encourages innovation and adoption of new ideas.**

Farmers place high credibility on demonstrations on nearby working farms. They perceive these as more relevant than either publications or experiment station demonstrations (Matheson 1989; Jefferson Davis Associates 1982). The farmer organizations have fostered wider acceptance of sustainable agriculture, a fact supported by their rising memberships. Cooperating extension agents working with these organizations also found that participatory on-farm programs are having an effect, as this Montana county agent notes:

> The funds you provided for us to start up this project are definitely making an impact on our people. There has definitely been more (leafy) spurge and noxious weed awareness come out of this project than I ever imagined. The chairman of the Weed Board said that the feature story in our local newspaper moved noxious weed control one step forward.

For innovative farmers, working in a group multiplies learning and reduces risk. For example, Farm and Ranch Improvement Clubs are an entry point into sustainable agriculture for Montana producers who previously may have felt isolated or unsure of where to start. The Land Stewardship Project found that group interaction encourages more thorough questioning of farm practices and enriches the investigation by subjecting assumptions to the critical scrutiny of others who see things differently.

All of the Initiative's studies found that a major benefit of participatory research was a greatly enriched opportunity for all to learn and apply new ideas. The process encourages learning exchanges among all participants and fosters respect for each person's contribution. To strengthen this interaction, the extension agent often changes roles from knowledge expert to group facilitator. *Technology transfer evolves into interactive inquiry.*

4. **Participatory research creates community and leadership development opportunities.**

All of the farmer-managed programs report that new relationships form and old ones become strengthened through participatory research. It creates a community for the farmers. The programs develop relationships, networking, and the opportunity to work with others in a positive experiment that offers hope for the future of the farming community. Considering the troubled times in rural communities, this hopeful alternative is welcome.

The leadership ability of farmers also blossomed through participating in the program. LSP found that farmers improved decision-making skills and self-confidence, as well as their financial and environmental condition. In a recent report, PFI relayed a result also found in the other programs:

> Many cooperators are now very comfortable speaking about their practices to groups of farmers and scientists. Farmer members of the board of directors have been especially challenged to carefully consider the broader issues of sustainability and the process of change in agriculture. Some are now serving on national boards, writing for local and national publications, and speaking to groups in other states and internationally. This has been "back door" leadership development. Skills have been developed out of specific needs and situations, always tied to the life situations these farmers face every day.

CONCLUSION

In summary, participatory research is a valuable tool, worthy of broader use in agricultural research and extension. Partnerships between producers and technical-assistance providers are leading to new ways of working. They are pioneering new directions and methods for research and extension. The partnerships speed innovation, adoption, and adaptation of new methods and technologies.

Perhaps most importantly, the community built by these programs is fostering new leaders and supporting the construction of a new future for rural communities, based on family farming and environmentally sound agriculture.

FOR FURTHER INFORMATION

Further information is available as indicated in each subsection of this chapter.

11

Case Studies and Decision Cases

Helene Murray
with contributions from
Melvin J. Stanford
Nancy Matheson

Whole-farm study is challenging because farms are complex entities. Decision cases and whole–farm case studies offer farmers and scientists alike insights into understanding the whole farm. These methods benefit (1) farmers, by identifying future research and extension priorities and involving farmers in agricultural research methods; (2) scientists, by demonstrating real-world farm complexity and how their expertise can be applied best; (3) laboratory-based researchers, by increasing their contact with farmers and extension personnel, which enhances university outreach; (4) students, by having them solve real farming problems; and (5) everyone, by improving communication about research among farmers, scientists, and extension personnel.

Case study methodology was largely refined in the business community, where it has been used to probe the complexity of economic forces that act upon an individual business. In agriculture, case studies have the potential to provide valuable information to supplement traditional agricultural research.

A case study can describe a farm or situation, examine issues and decisions to be made, illustrate a phenomenon, be an educational tool, and foster interdisciplinary work. Case studies provide an excellent avenue for enhancing a scientist's knowledge of real-world farming. This method also creates a comfortable framework that fosters collaboration between scientist and farmer.

A whole-farm case study (WFCS) is a systematic examination over time of the biological, social, and economic factors of an entire farming system. It examines interactions among production practices, economic status, business management, and relation of the farm family and employees. Because WFCSs are designed and conducted to understand entire systems, they are best conducted by interdisciplinary teams representing both the biological and social sciences.

As farmers participate in a whole–farm case study, they:

- Describe their farming practices.
- Describe their management strategies.
- Describe innovations developed on-farm.
- Describe opportunities they have for making changes, and the constraints.
- Help identify areas that need additional study.

The very process of conducting whole-farm case studies helps forge new working relationships among farmers, researchers, and extension personnel, thus encouraging participatory efforts (Chapter 10).

A decision case is distinct from the WFCS because it is primarily a teaching tool that engages participants in problem-solving exercises. A decision case is a documentation (sometimes written, sometimes video-taped) of a real problem in a real farming situation. Decision cases provide the following information (Stanford et al. 1992):

1. An identified decision-maker.
2. Issues involved in the situation.
3. Objectives of the decision-maker.
4. An outline of feasible alternatives.
5. Information required for analysis, support, appraisal, and background.
6. Work for participants to perform.

Both whole-farm case studies and decision cases were used in the Foundation's Initiative to better describe and understand agriculture at the farm level, and to identify research and education needs.

FARMERS SPEAK THROUGH WHOLE-FARM CASE STUDIES (WFCS)

In a 1987 study by the Alternative Energy Resources Organization (AERO), farmers in arid regions of the Northwest identified their preferred sources of information about sustainable farming practices. They ranked *other farmers* first and *farm tours* second (Matheson 1989).

If whole-farm case studies are viewed as "scientific farm tours," then the published reports from WFCSs are "arm-chair farm tours." Thus, these reports offer a useful substitute for first-hand farm visits, and are valuable in education and diffusion of sustainable practices. WFCSs also provide farmers, researchers, and anyone interested in agricultural systems with a model for identifying innovations.

Nine WFCSs in the Semiarid Northern Plains

Nine WFCSs conducted by AERO in the dryland Northern Plains, Canadian Prairie, and Intermountain Northwest examined *sustainable farming systems that reflected the use of nature as a model for agriculture* (Matheson et al. 1991). Each farm used practices geared toward the ecological and climatic conditions at its location. AERO documented the time-tested understanding that these farmers had of how nutrient and pest cycles maintain fertility and minimize pest occurrence or damage.

The nine farms in the study demonstrated how sustainable cropping systems function in semiarid regions. Each farm had unique constraints, and each farmer possessed a unique palette of talents and resources. Yet, their *general situations* were not unique. Their farm size, financial condition, and environmental characteristics matched the profile typical of many farms in the region.

The farmers described in the study have successfully implemented innovative crop rotations that are productive, profitable, and which meet their objectives for reducing potential environmental and health problems, building soil, diversifying income potential, and controlling weeds and other pests. *The usefulness of these WFCSs lies in presenting their successful methods so that other farmers and scientists can adapt them to other farming situations.*

WFCSs of Oregon and Washington Horticultural Farms

A joint Oregon State University/Washington State University study in western Oregon and Washington used WFCSs as an early stage of a larger program designed to increase farmer participation in research and extension (Figure 11-1). The process shown illustrates a model for combining farmer-scientist knowledge through several phases. The Oregon/Washington team found WFCSs valuable during certain problem-solving stages, and invoked other methods to complement the case studies when necessary (Murray et al. 1994b).

The Oregon/Washington study examined 16 farms that produced vegetables and small fruit. They ranged in size from 8 to 3,000 acres and displayed a range of production systems, including certified organic farms. In addition to the 16 farm families, the Oregon/Washington team consisted of 34 researchers and extension staff.

The process of conducting WFCS research provided a rare opportunity to work with other disciplines on farms and provided team members with new perspectives. The vast majority of university participants stated that they would consider using the WFCS approach again in their work. However, the primary constraint cited by all team members was the time required to conduct the study.

In western Oregon and Washington, the process of conducting WFCSs led to a better understanding of farming systems by the researchers and extension agents on the team, identified areas requiring study, and initiated longer-term projects involving joint farmer-scientist participation.

For example, because of the success of these case studies, a coordinator has been hired in Oregon to assist farmers in conducting on-farm research. This individual provides information about managing trials and evaluating results, while the farmers define research topics and manage the trials on their own farms.

Another development resulting from the WFCSs has been scientist-managed experiment-station research to complement information gained through on-farm research. For example, studies in cover cropping, beneficial insects, and nitrogen management now are being conducted on experiment stations and on-farm by both farmers and scientists as a result of the whole-farm case study project in western Oregon and Washington. (Please see Sidebar 11–1, "A Farm Tour in Print: The Case Study.")

SIDEBAR 11-1

A Farm Tour in Print: The Case Study

(Adapted from Helene Murray, Richard Dick, Daniel Green-McGrath, Lorna Michael Butler, Larry S. Lev, and Richard Carkner, Whole Farm Case Studies of Horticultural Crop Producers in the Maritime Pacific Northwest, 1994b. The editor thanks Oregon State University for permission to use material from its report.)

As farmers respond to changing farm policy and practices, they are experimenting with new methods of production and farm management. The most practical information on these new methods exists where the planter meets the soil—right on the farm. Case studies are an open gate through which to access the discoveries of these innovators.

WFCS Model from Oregon/Washington

During 1989-1990, Oregon State University and Washington State University conducted whole-farm case studies of 16 vegetable and fruit farms across a 300-mile stretch of western Oregon and Washington State.

They sought to better understand these farms and their operators, to help target research and education for the region, and to discover alternative approaches to solving problems.

OSU/WSU teamed two dozen diverse research and extension specialists in the biological and social sciences: agronomy, economics, ecology, family studies, home economics, soils, farming systems, marketing, horticulture, anthropology, and entomology. The team identified 25 organic and progressive conventional farms as case-study candidates, based on their significant use of innovative practices. The team then conducted a brief survey called a *sondeo* (Spanish: sounding or fathoming) to find farmers interested in participating in a longer–term case study.

Small sondeo teams (usually the county extension agent, a social scientist, and a biological scientist) informally interviewed candidates: What crops do you grow? How do you handle pest problems? Do you farm full time, or work off-farm too, and do you hire labor? What new practices have you tried, and how well do they work? How are they different from your neighbors'? How do you market and keep records? Are you in a grower organization? What is your farm's toughest problem? Will you participate in our case study?

The team selected 16 farms for case studies, ranging from 8 to 3,000 acres, with annual gross sales from $10,000 to $4 million. They included seven organic producers, eight conventionals, and one with both types of production. Data were collected as field notes from farm tours and interviews (informal and structured), farmer-completed forms on monthly labor and fuel use, soil and plant test results, photos of production practices, and any articles in the popular press on the farms and families.

Each farm was visited by the project coordinator and others at least three times. Visit #1 (June-July) focused on production practices, equipment, decision-making, marketing, and information sources. Visit #2 (August-September) examined production practices and decision-making during the peak growing season. A topic list guided the flexible interviews:
• Farm/household description (size, cropping history, soils, marketing, family profile, perceived strengths/weaknesses).
• Production practices (rotation, livestock management, pest control, equipment, information sources, perceived problems).
• Socioeconomic information (family member roles, off-farm employment, commitment, goals, organizational involvement, land ownership, capital sources, record-keeping, insurance, etc.).

Visit #3 (November-February) asked open-ended questions on management, economic status, and labor issues. In all, 43 people were interviewed at the 16 farms. After each group of visits, team members compared findings and impressions.

At the project's midpoint, a half-day Farmers' Forum was held to introduce all participants, share ideas and experiences, receive feedback on preliminary findings, and prioritize future research and education activities. Attendees from 13 of the 16 farms expressed particular interest in sharing ideas and shaping future research.

Case Study Results

These whole-farm case studies yielded striking portraits of the farms. The results include many elements, including commitment to farm life, soil analysis, weed and pest management, plant disease, family participation, production practices, farm succession, finding useful information on sustainable farming, farm labor, wages, regulations, worker documentation, incorporating computers into management, land, capital sources, insurance, developing farm–level value-added opportunities, and promoting public interest in agricultural policy.

Collectively, the case studies disclosed the top concerns of growers: hired labor problems, banning of some pesticides, rising production cost compared to nonrising farm-gate prices, insufficient land, weather, urban sprawl, pesticide drift from conventional farms onto organic farms, and pest management. The overall project report (Murray et al. 1994b), with its candid quotes from farmers and researchers, offers the reader an intimate window on the activities of case–study participants and how they resolve their respective problems.

With a few dissenters, participants highly valued the case study experience. Mingling farmers with members of several disciplines produced a powerful hands–on learning experience. Clearly, much of the information gained from these real-world case studies is unavailable in laboratories or at experiment stations, demonstrating the value of the case-study technique.

Of course, no two farms have the same resources, limitations, and management options, so these case studies cannot be recipes for success. They are models to learn from. To facilitate the conduct of whole-farm case studies, a case–study manual is available (see For Further Information).

The Value of WFCSs

WFCSs proved to be a useful framework for organizing research and extension at the farm scale, when certain criteria were met:

1. *Complementary skills* must be brought to joint research and extension efforts by the farmers and scientists.
2. *Communication among all study participants* is essential regarding study design and expectations.
3. *Flexibility and incremental design* are essential for joint projects. Participants must be willing to revise plans and strategies and to maintain a balance between the needs of observer and observed.

In summary, whole-farm case studies improve communication among varied people, when conducted properly. Farmers benefit by identifying priorities for future research and extension. They also learn more about agricultural research strategies and methods. Scientists develop a better grasp of farm complexity and how their expertise can best be applied. Farmers also report that they gain new insights and ideas about their own farms and operations by answering questions posed by people who are not directly affiliated with their farms.

Because problems do not occur in isolation, an understanding of entire farming systems helps put the interactions into context. Whole-farm case studies enable laboratory-based researchers to visit farms and increase contact with farmers and extension personnel. WFCSs are a way to enhance university outreach.

TEACHING WITH DECISION CASES

Decision cases are a valuable decision-making tool for situations that are broad, complex, or ambiguous, and where no clear-cut optimal, democratic, or rigorously rational solution may exist. This describes many agricultural decisions, so they may be broached effectively by the decision-case method. Today we hear calls to incorporate farm-level concerns (human, ecological, and social factors) into research and education, and decision cases offer a process to address these complex issues.

Work conducted by the College of Agriculture at the University of Minnesota during 1988-1993 demonstrates the value of decision-case methods in agricultural research and teaching. Of more than 40 decision cases prepared at the University of Minnesota, 20 relate to issues in sustainable agriculture, such as soil conservation compliance, crop rotation, and farm succession (what happens to the farm when the farmer retires or dies). These 20 cases provide unique information about sustainability in agriculture.

The decision-case method identifies issues, constraints, options, and other information from the viewpoint of the decision-maker, usually the farmer. Decision cases now are a key component of undergraduate and graduate agriculture courses at the university, affording a better understanding of farming system complexity. A number of other universities are experimenting with decision cases in a variety of agriculture classes—for example, Michigan State, Purdue, Texas A&M, and Utah State.

Decision cases also are being used in extension to add real-world context for specific problems. For example, decision cases have been used to teach farmers how to apply scientific principles in problem-solving. They provide a powerful new teaching strategy which could be widely adopted in extension.

For sustainable agriculture issues, we found that decision cases—

- Provide a better-integrated understanding of agricultural, social, and personal issues in farming operations.

- Organize information about complex and uncertain issues.

- Provide information about practices and systems that may identify areas for further study.

- Facilitate cooperative research between scientists and farmers.

- Provide teaching materials that place students in the role of the farmer or decision-maker and include current issues in applied situations.

- Present an avenue for future agricultural research and education in sustainable agriculture.

Please see Sidebar 11–2, "Decision Case Example: A Problem Child."

SIDEBAR 11-2

Decision Case Example: A Problem Child

(Adapted from R. Kent Crookston and Melvin J. Stanford, Dick and Sharon Thompson's "Problem Child," a Decision Case in Sustainable Agriculture, 1990. Copyright 1990 by R. Kent Crookston, University of Minnesota, and Melvin J. Stanford, Mankato State University. This decision case was prepared as a basis for class discussion and research. For further information, contact R. Kent Crookston, Department of Agronomy and Plant Genetics, University of Minnesota, St. Paul, MN 55108. The editor thanks the authors for permission to use material from their decision case.)

To illustrate a decision case about an issue in sustainable agriculture, here is an example from an innovative Iowa farm. It has been greatly simplified from the actual decision case used in agriculture classes at the University of Minnesota and other schools.

Plants require three primary nutrients for good nutrition: nitrogen, phosphorous, and potassium. An important task for any crop farmer is to provide the proper amount of each nutrient at the proper time. Conventional farmers often add these nutrients by treating fields with chemical fertilizers. In contrast, sustainable farmers do their best to find non-manufactured ways of doing so, such as addition of manure or growing cover crops that contribute nitrogen to the soil.

About 30 years ago, Dick and Sharon Thompson stopped using chemical pesticides and fertilizers on their 300-acre farm near Boone, Iowa. The farm was average in size, but extraordinarily diversified. The typical crop rotation in the area is corn and soybeans, but the Thompson's unusual five-year rotation of corn-soybeans-corn-oats-hay comprised all of Iowa's principal crops on a single farm. In 1989, the National Academy

---

of Sciences featured their innovative farm in its landmark report, *Alternative Agriculture.*

The Thompsons, outspoken advocates of sustainable farming and especially ridge tillage, held field days for visitors to "see . . . how a system based on cover crops, manure, and ridge tillage can save money and soil, increase infiltration [of water into the soil], improve soil tilth, fix and hold nitrogen, control weeds, and stimulate earthworm activity."

Like all farms, this one had its problems. Most persistent was a potassium deficiency affecting the corn, so much so that the Thompsons called potassium deficiency their "problem child." What was causing it?

Thompson carefully reviewed every aspect of the operation that might affect potassium levels. He analyzed his handling of manure (a potassium source) and how it was stored and spread on his fields. He studied years of soil and leaf-tissue analyses for potassium levels. Thompson found that he had achieved excellent potassium levels in the soil, and had even applied more than needed. Yet, leaf-tissue tests showed too little *uptake* of potassium into the plants. Something was inhibiting potassium assimilation into the corn.

Thompson was not alone. Other farmers who practiced ridge tillage also were experiencing potassium deficiency. At a meeting of the Practical Farmers of Iowa, Thompson saw a report in which researchers suggested that tillage practices and drought were the culprits.

But Thompson suspected otherwise. An Iowa State University agronomist had experimented on some of Thompson's acreage, adding *chemical* nitrogen and getting 179 bushels of corn per acre. Right beside this plot, Thompson had used *manure* for nitrogen, and gotten only 141 bushels per acre. Could his use of "natural nitrogen" from manure somehow be affecting potassium uptake?

Included with the full decision case are numerous data exhibits on the Thompson farm. These allow students to investigate the problem in depth and seek clues. The decision case then asks students these questions:

• What do you think caused the potassium deficiency? How could you confirm it?

• Evaluate Thompson's observation: "We may have discovered the answer to the potassium uptake problem but didn't know it. It may take some liquid potassium under the seed, along with dry potassium off to the side, to reach early potassium uptake in the plant."

• For almost 20 years, the Thompson farm had been certified as an "organic" farm, meaning that they agreed not use chemical fertilizers. But the Thompsons had no evidence that the chemical fertilizer containing potassium was a problem to the environment. So they started using it, improving yield but losing their "organic" status. Evaluate their reasons, and decide how you think it affected their business and the environment.

• How would you evaluate the economics of their potassium management?

• How relevant is Thompson's experience to other farmers?

> This example follows the typical pattern of a decision case: it identifies a decision-maker; cites the issues involved; indicates the decision-maker's objectives; outlines feasible alternatives; presents information required for analysis, support, appraisal, and background; and provides work for participants to perform (questions to be answered).
>
> (To obtain a complete copy of this decision case and a list of others available, see For Further Information.)

CONCLUSION

Whole–farm case studies and decision cases offer a systematic means of compiling information in complicated areas of human endeavor, providing useful observations that go beyond the range of controlled experiments. Individual components of farms can be compared statistically—for example, crop varieties, livestock rations, or productivity. However, whole-farm studies provide insights into how things work and interact in "real life" settings and may reveal what traditional agricultural research cannot. When used appropriately, case studies offer new insights to both farmers and scientists, enabling them to better address system–wide issues, problems, and possible solutions.

FOR FURTHER INFORMATION

These resources provide information about whole–farm case studies and farmer/scientist focus sessions:

Matheson, Nancy, Barbara Rusmore, James R. Sims, Michael Spengler, and E. L. Michalson. 1991. *Cereal-Legume Cropping Systems: Nine Farm Case Studies in the Dryland Northern Plains, Canadian Prairies, and Intermountain Northwest.* Alternative Energy Resources Organization (AERO), 44 North Last Chance Gulch #9, Helena, MT 59601 (406/443-7272).

Murray, Helene, Larry S. Lev, Daniel Green-McGrath, and Alice Mills Morrow. 1994. *Whole-Farm Case Studies: A How-To Manual.* Oregon State University Extension Service EM 8558. Publications Orders, Agricultural Communications, Oregon State University, Administrative Services A422, Corvallis, OR 97331-2119 (503/737-2513).

Murray, Helene, Richard Dick, Daniel Green-McGrath, Lorna Michael Butler, Larry S. Lev, and Richard Carkner. 1994. *Whole–Farm Case Studies of Horticultural Crop Producers in the Maritime Pacific Northwest.* Oregon State University Station Bulletin 678. (Same ordering address)

To obtain more information about decision cases, a list of those available, and their associated teaching notes, please contact: Program for Decision Cases, University of Minnesota, 411 Borlaug Hall, St. Paul, MN 55108, 612/625- 7773.

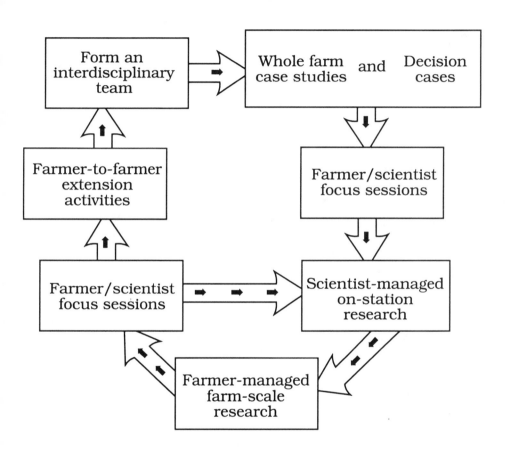

Figure 11-1. One model for combining approaches to increasing farmer participation in research and extension.

12

Using Ecological Methods in Agriculture

John C. Gardner

The science of ecology offered essential concepts and methods for our studies, for a farm is an ecosystem. Ecology lent a framework for relating observations at different scales (for example, relating field observations to whole–farm observations). Ecology provided the language to discuss interactions and identify each enterprise's role in a whole farm. Ecological methods provided a benchmark for comparing farms and a goal for reducing agriculture's environmental impact compared to the natural grassland climax ecosystem.

Our search for methods to study farms and their associated communities would not be complete without borrowing from the science of ecology. It is perhaps the only discipline among the natural sciences explicitly concerned with the *importance of scale* in its study of the relations between living organisms and their environment. We used ecological concepts, such as *ecosystem* and *niche*, as a framework for conducting comparative studies among farms to assess their impact on the environment.

STUDYING THE WHOLE AS WELL AS THE PARTS

A central objective of the Initiative was to study agriculture, not at the field level, but from a *whole-farm perspective* and a *farming-community perspective*. This larger scale presented an immediate hurdle. Methodologies used in agriculture are scaled to studying individual parts of the farm, such as water and nutrient flow through specific soil sites, or the impact of tillage on soil properties over a season, or the nutritional requirements of gestating beef cattle. In most agricultural literature, the overall impact is merely the sum of these individual parts, with little consideration of their interactions.

But we found ample evidence that farm management influences the biological characteristics of the whole farm, and that certain relations seemed consistent, regardless of locale. The management strategy of greater crop diversity, for example, nearly always was linked with a greater diversity of weed, insect, and disease species. In many cases, this diversity helped the farmer avoid devastating crop losses and reduced the requirement for outside input.

By gathering data on the whole farm, and not just its parts, we clearly saw the impact of farm management. Results sometimes ran counter to the intuitive conclusion that would be reached from just a traditional "summing of the parts." Here are two examples:

- Eastern North Dakota receives greater annual rainfall than the more arid western part. Therefore, one would expect greater potential for leaching by soil water to be in the higher-rainfall east. However, we found the opposite on certain farms. The *whole-farm view* revealed why: it was possible for a western no–till system of soil conservation to retain more water than could be held in the soil profile.

- In the northern Plains, the common "wheat/fallow farm" alternates spring wheat and fallow year-to-year. Because wheat is grown one year and the soil lies fallow the next year, one would expect the soil to have less organic matter than a farm that plants crops every year. With the soil unprotected by plant cover every other year, and with less organic matter to "glue" the soil together, one would expect the soil to be more erosion-prone. However, we found the opposite on one wheat/fallow farm. Again, the *whole-farm view* revealed why: the farm included a sizable beef cattle enterprise that employed cyclic spreading of the animal manure among the fields.

Although we found repeated quantifiable evidence of such whole-system characteristics, we had difficulty assembling the data in a manner that the agricultural literature would find credible. Methods were needed to rigorously test, interpret, and share these important interactions at the farm level (landscape scale).

Fortunately, such methods were available. Among the great conceptual contributions to come from ecological science are the *system* (E. P. Odum 1969; H. T. Odum 1983) and *hierarchy theory* as applied to ecology (Allen and Hoekstra 1992). Ecologists have described numerous examples of ecosystems, and have quantified them. Based on these, it is not much of a leap to suggest that *a farm is an ecosystem of its own.*

Thus, we needed the systems science developed by the discipline of ecology. However, we were well aware that this was no panacea, for ecologists themselves are divided on the validity of systems–science concepts (Mansson and McGlade 1993; Patten 1993). If agriculture is to adopt the thinking and methods of system ecology, it is wise to become familiar with some of its critics.

But, despite the debate—perhaps because of it—the merging of agriculture's practical and empirical bent with the theoretical basis found in systems and hierarchy ecology promises to be at the forefront of the development of more sustainable agricultural production systems (Coleman 1989; Jackson and Piper 1989; Paul and Robertson 1989; Elliot and Cole 1989).

Other whole-farm findings reinforced the value of using the larger-scale ecological-systems approach. We found that single-objective farm-management practices often resulted in a broader environmental impact than intended. In North Dakota, a study that compared native prairie sites with conventional, organic, and no-till farming systems provided these interesting examples:

- Agricultural soils differed from prairie soils most in the east, and least in the west. Organic matter was found to be depleted by 71% in the east, 57% in central North Dakota, and 27% in the west.

- Two common types of fallow are "black" fallow (weeds are controlled by tillage, leaving the soil black) and "brown" fallow (herbicide is sprayed, leaving brown dead plants that hold the soil and provide cover). Moisture storage is much greater with brown or "chemical" fallow (23%) than black fallow (5%), because brown fallow permits less evaporation. However, this greater moisture in the soil created more water movement. Thus, although brown "chemical" fallow helped reduce soil erosion, the increased water movement may carry the contaminants into groundwater, threatening its quality.

- The crop diversity and surface cover of organic and no-till sites increased the numbers and kinds of insects found. However, among them were many beneficial insects. Large populations of ground beetles at organic sites may play a role in reducing the number of weed seeds.

- Among the agricultural sites, overall soil-quality characteristics were best where tillage was minimized, or where organic matter returned to the soil was increased through animal or green manures.

FINDING AN ENVIRONMENTAL BENCHMARK

One of the most spirited debates surrounding contemporary environmental issues is that of "appropriate" or "best" land use. Most view a "natural" area as the best example of environmental integrity. However, many natural areas far exceed farmland in soil erosion. Also, nature sometimes produces less diversity than human systems (Loucks 1970). Regardless, we needed an experimental control against which to ecologically compare whole-farm systems. We found no better benchmark than that of the *natural climax ecosystem* as prescribed by the science of ecology.

A natural climax ecosystem is the end-product of natural processes. It results from *succession,* one of the most studied ecological concepts. Succession refers to the natural progression of organism types in an environment over time, with some types out-competing others until the most dominant ones attain full stature. This eventual product of an environment, with its plant and animal organisms interacting over time, is the *climax ecosystem* or *steady-state ecosystem.* A textbook example is the deciduous forest, which begins with grasses, then shrubs, then small trees, then progressively larger and longer-lived trees, until a climax forest of towering hardwoods is attained.

Ecologically speaking, most midwestern agriculture resides in a perennial grassland. In other words, this grassland *is* the climax ecosystem, because conditions do not permit trees to compete effectively. Its soils are the product of millennia of perennial grasses interacting with a host of herbivores, carnivores, omnivores, climate, and prairie fires, all of which helped the grassland evolve.

In such an ecosystem, the majority of plant tissue, nutrients, and stored energy is below ground, in the roots. But our agricultural systems harvest the portion above ground, so farmers are in annual combat with the nature of a perennial plant community.

Such prairie ecosystems were a benchmark not only within a region but across regions. Because these natural climax ecosystems were independent of current agricultural practices, they also served as just the experimental control sites we needed to eliminate bias when comparing farm management practices. As might be expected, farms that emphasized the conservation and cycling of organic matter through the soil system were those that most closely mimicked the natural prairie.

206

Others have suggested that the natural climax ecosystem has even further value as an agricultural concept. Jackson (1985) suggested that the best approach would be to make agro-ecosystems be *analogs* of natural systems. For example, in the prairie of the Great Plains, the dominant crop plants should be *perennial* grains, just like the natural grasslands evolved over millennia, and not *annual* grains. (Perennial plants have a life cycle that continues for years; annual plants live only through one growing season and persist only through annual production of seeds.) Jackson's thoughts have stimulated a whole area of research toward such new agro-ecosystems (Soule and Piper 1992).

Most agree that agricultural analogs of natural systems would move agriculture toward more ecological sustainability. Some feel that Jackson envisions the wrong *role* for the crop plants, however, arguing that perennials are for grazing, not for grain (Vogel 1992).

Knowing the ecological players and processes in natural ecosystems seems destined to play a larger role in agriculture. Ecology has the theoretical framework, while agriculture has the practical experience. Now both seem willing and able to contribute to the development of productive new agro-ecosystems at less environmental cost.

CONCLUSION

The science of ecology offered many useful concepts to the Initiative. Practically, it provided a framework in which to relate observations taken at one scale (such as a field) to that of another (such as a farm). It provided the language to discuss interactions and to identify the role that each agricultural enterprise played in a whole-farm setting.

Conceptually, it provided an ecological benchmark for comparison among farms and regions, through use of natural climax ecosystems. Such native systems were used not only as comparative benchmarks, but also as goals in reducing the environmental impact of agriculture. Though our use of ecological concepts was limited to biophysical properties of agriculture, other researchers have extended these concepts into economic and social-issue analysis (such as Daly and Cobb 1989). As disciplinary thought and study further converge as applied to agriculture, the science of ecology seems ready to contribute.

FOR FURTHER INFORMATION

More information on the farm/prairie comparison study in North Dakota is available in the report *Farming Practices for a Sustainable Agriculture in North Dakota* (Clancy et al. 1993).

13

Geographic Information Systems (GIS)

Mary Lynn Roush

A Geographic Information System (GIS) allows simultaneous computer analysis of multiple kinds of information about a locale. Maps generated with a GIS are used to study "what-if?" scenarios, such as drought. GIS-mapped ecological and sociological data will be used to project land-use trajectories with greater certainty. GIS is moving us toward implementing the "conceptual model" approach for addressing key resource issues about a region.

Another approach to evaluating the impact of agriculture across various scales is emerging with the advance of computer technology and information processing. A **Geographic Information System (GIS)** allows simultaneous computer analysis of multiple kinds of information about a locale—biological, ecological, agricultural, soils, climatic, topographic, or demographic.

A GIS electronically overlays the parameters one selects, and the result is examined for patterns and combined effects. For example, overlaying crops, soils, and demography for a county in Oregon allows one to see the patterns of cropping and communities in relation to soil types. Trends of cropping and settlement can be projected.

GIS: THE OREGON EXPERIENCE

In western Oregon, an interdisciplinary team of agricultural scientists, ecologists, social scientists, and farmers used a GIS to identify ecological and social forces that influence the sustainability of land-use patterns and cropping systems. We looked for changes in land-use patterns, cropping systems, and regional socioeconomic infrastructure as indicators of how farm-scale agricultural practices relate to national-scale policies.

To explore some preliminary scenarios, we used a powerful GIS named ARCINFO to organize a database for the Willamette Valley. Two maps were generated:

1. An *ecological* map of soil types and associated environmental information.
2. A *social* map, with county boundaries, containing socioeconomic and cropping-system information from the 1987 agricultural census.

We used these maps to study "what-if?" scenarios. We tried varying ecological factors, such as water limitation, to see the impact on farm areas. We also varied social factors, such as urban encroachment onto rural land, to view the impact.

A GIS EXAMPLE: DROUGHT, PLANT FITNESS, AND WATER ALLOCATION

Water management not only is important in sustainable agriculture; it also is a good illustration of GIS utility as an analytical tool. Let us consider the impact of water allocation in Oregon, a state that, not long ago, felt immune to water shortage. The past decade dramatically changed that perception:

- After six years of prolonged drought, Oregon reservoirs are dangerously low. Many water districts face potential loss of irrigation water.

- Oregon is under increasing pressure from California to divert water southward.

- Continuing drought and water-quality problems in California may open a new market there for Oregon agriculture. For example, poor water quality in California's lettuce-growing areas may increase demand for Oregon lettuce.

We used the GIS to integrate information on regional soils and to explore the implications of water limitation for Oregon agriculture. First, we calculated a *fitness index* for each Willamette Valley crop, indicating how effectively it converts natural light and water into biomass without supplemental irrigation or drainage. In essence, this index shows each crop's sensitivity to drought. A high-fitness crop uses available water and nutrients efficiently and does not require a lot of input (water, fertilizer).

Low-fitness crops (strawberries and row-crop vegetables such as broccoli and sweet corn) are immediately jeopardized by a drought. However, because they usually are irrigated, low-fitness crops are affected least by a mild or brief drought. If drought worsens, it is the

210

high-fitness *nonirrigated* crops that are first to suffer, such as winter wheat and grass seed. Because irrigated land is less sensitive to moderate drought, farming systems tend to move toward low-fitness/high-value irrigated crops if drought continues.

The paradox is that reservoirs and whole watersheds may face water limitations that coincide with increasing dependence on irrigation. After six years of prolonged drought, Oregon faces abrupt loss of irrigation water over large areas if snowmelt and rainfall fail to raise reservoir levels. Water management districts may be forced to recall water rights that have been granted to users.

Crops and cropping systems with low fitness are less sensitive to deteriorating soil quality, at least for awhile. With these crops, farmers may have the impression that they can "get away with" practices that damage the soil resource. Such practices include high-intensity fertilizing, irrigation, and tillage, repeated year after year instead of alternating with years of reduced intensity.

The impact on soil health seems subtle as long as energy, water, and capital are plentiful. But when these resources grow scarce, crop production becomes highly dependent on soil condition and on good agronomic practice (Joy Tivy 1990, personal communication).

OTHER GIS-BASED RESEARCH PLANNED

With more data added to the GIS, further "what-if?" questions could be addressed:

- If Oregon's drought ends and California's continues, can Oregon effectively take over a larger share of the produce market?

- Does the infrastructure of Oregon agriculture help or hinder its ability to respond to new markets?

- If Oregon's drought continues and irrigation rights are curtailed, which counties will be hardest hit?

- In which counties can farmers respond by changing cropping systems to more efficient crops that demand less water, such as grass seed and winter wheat?

- How will the current systems of markets and processor contracts impact the flexibility of farmers?

- What might be the long-term effects of shifts in cropping systems for Oregon's agricultural markets and farm communities?

We are developing the GIS method to explore policy implications of other natural resource issues in the Pacific Northwest, particularly forestry. Using diverse data sources (local, state, national, and private), maps, and remotely sensed imagery, we are constructing land-use changes over time. For some areas, we are analyzing actual vegetation, cropping systems, and land uses from maps; for others, we are reconstructing this information from data.

Next, we will explore mapped ecological and sociological data to *explain* changes in land use. And from this, we will *project* land-use trajectories with greater certainty than ever before. GIS is moving us toward implementing the "conceptual model" approach for addressing key resource issues in the region.

14

Overcoming Barriers: Reflections on Cooperative Research

Karl N. Stauber

When the Initiative began, university researchers and alternative agriculture groups "knew" they could not work together. By the end of the effort, these people learned they could work together. Today, several academic institutions involved in the project have created mechanisms for working across disciplines and institutions. In three short years, alternative agriculture groups in these states have gone from being outsiders to being important research allies.

The Northwest Area Foundation's goal in the Sustainable Agriculture Initiative was to inform the public policy debate on the future of agriculture. To achieve this, the Foundation believed it was critical to draw upon two valuable sources: *the scientific expertise of university researchers and the practical experience of farmers.* Each represented an unduplicated set of strengths and capacities needed for the project's overall success.

Consequently, the Foundation designed the Initiative's program to support *cooperative research, conducted jointly by universities and alternative agriculture groups.* However, what began as a straightforward effort to fund research quickly became an exercise in multiculturalism. Without realizing it in advance, the Foundation had wandered onto the construction site of the Tower of Babel.

University researchers questioned the role farmers ought to play in the research process. Several researchers thought farmers should serve only as the *subject* of research. However, because farm practices and their consequences were the focus of the project, the Foundation believed that living, breathing, dirt-under-the-nails farmers had to play a central role. The Foundation insisted that farmers should have a voice equal to the researchers in designing and conducting the research.

The farmer-controlled alternative agriculture groups came to the table with similar limiting assumptions. Because of past negative experiences, they were distrustful of some university personnel. These groups often viewed the university as a single entity: if farmers had a problem with one part of the institution, they would have the same difficulty throughout the institution. In addition, some thought that the research conducted should be designed primarily to meet their particular needs. Research that did not directly benefit farmers was not worth doing. Finally, farmers wanted answers to their problems quickly—this year, not years from now. Most research, however, cannot provide useful information on such a timetable.

The Foundation's funding offer brought both groups to the table. But even as the project was in planning, some researchers and farmers questioned whether they could work together. Several academic administrators approached the Foundation and asked that it remove the alternative agriculture groups from the discussion. "We can do this better without them," they assured the Foundation. One alternative agriculture group indicated that it would not have its agenda dictated by outsiders. What was the Foundation to do?

With much help from consultants and advisors, the Foundation made a decision that, in retrospect, was on-target and established a model for future cooperative research: *continue to have the two groups work together, but in a manner that would allow for new relationships to evolve, and specifically help the two groups invent a way to work together.* Each group was given support to build its own organization, as well as money for the Foundation-initiated project.

Even within the university setting, problems surfaced. To the Foundation's surprise, most of the universities it approached had no mechanism in place to support and reward *multidisciplinary* research. Although individual faculty at every institution wanted to participate, they felt they would be penalized for their involvement. This raised serious questions about existing requirements for tenure and promotion, and whether they effectively prohibit much useful research.

Once researchers were engaged in the project, the lack of a common language presented another unanticipated hurdle. At the first meeting of project participants, an agronomist challenged a sociologist: "You're not talking about research! If it cannot be replicated, it's not real research!"

When the project began, the various groups "knew" there were reasons they could not work together. By the end of the effort, people "knew" there were reasons and ways they *could* work together. Today, a number of the academic institutions involved in the project have created mechanisms for working across disciplines and institutions. In three short years, alternative agriculture groups in these states have gone from being outsiders to being important research allies. These changes were not caused by the Foundation's research project, but they were helped by it.

There always have been reasons to work together and reasons to stay apart; probably there always will be. But now there is a public model that makes "it can't be done" a false statement.

SIDBAR 14-1

Keys to Successful Cooperative Research

Karl N. Stauber

If there is any undertaking that demands perception, sensitivity, and dynamic management, it is cooperative research. These keys emerged from our experience:

• *Seek what is common among participants.* At every meeting, *acknowledge what is common and unifying.*

Both the academics and the farmers cared about agriculture and their communities. This mutual concern committed them to the project from the start. Over time, the participants discovered other common ground. Although social scientists and physical scientists had different ideas about research, both believed in the rigorous search for knowledge. While farmers wanted answers today and researchers were willing to wait until next year, they sought answers to similar questions. Finally, everyone came to see the project as an exercise in building community and doing research. These objectives clearly were mutually supportive, not contradictory.

• *Create opportunities for cooperation,* because they will not occur on their own.

The entire Initiative created opportunities for cooperation—grant money it provided enabled cooperation to happen. This book, with its thirty-some diverse contributors from different disciplines and groups, is an example of such cooperation. Cooperation across institutional and disciplinary lines does not come easily, so the opportunities must be created—and seized.

• *Cooperation is a learned behavior that can be taught and assisted.*

Help from outsiders can be critical in clarifying roles within a group. The Center for Rural Affairs played a critical role in this regard. It consistently "modeled" and promoted cooperation.

• *Respect the advantages your differences bring.*

Recognize what benefits each group brings to the table and how they can enhance the project objective. Give credit where credit is due. Academics need to recognize and support farmers' management expertise; farmers

---

215

need to recognize academics' resources for information collection and analysis.

• *Learn to tolerate the inefficiency that differences may bring to the project.* Farmers had to tolerate the bureaucratic inefficiency of research institutions. Academics had to tolerate the informality of small nonprofit organizations.

• How *something is said may be as important as* what *is said.* Pay attention to *the meanings that each group assigns to the words it uses.* Remember that language creates opportunities for both exclusion or inclusion. At one field day, a farmer and a researcher argued over whether the farmer's application rate of different chemicals to his fields was significant. The researcher suggested that the difference in application rate was not significant. The farmer, visibly angered by this comment, felt that the researcher was demeaning his work. But the researcher was talking about *statistical significance.* The farmer was talking about *economic and environmental benefit.*

• *Above all, remember to be patient. New relationships take time to develop.*

PART

IV

FUTURE RESEARCH NEEDS AND POLICY ISSUES/RECOMMENDATIONS

The results of the Northwest Area Foundation's Sustainable Agriculture Initiative are only a beginning toward answering many questions about sustainable agriculture, rural communities, and innovative approaches to agricultural research. Chapter 15, "Future Research Needs," reviews some of the remaining questions, in four categories: (1) defining and identifying sustainable systems, (2) economics of sustainable farming practices, (3) implications for social structure of a shift to alternative farming systems, and (4) conditions that shape farmers' decision-making. Needs for policy and institutional research are included.

Chapter 16 concludes the book, reviewing the findings reported in Parts II and III and considering their policy implications for farming, rural development, and agricultural marketing, environment, research, and extension education. New directions are suggested for institutional and legislative policy to improve the potential of sustainable farming. A sidebar analyzes alternative farm policy options that were debated in the 1990 Farm Bill, and which remain relevant to the 1995 Farm Bill.

15

Future Research Needs

Elizabeth Ann R. Bird
with Brenda J. Johnson

and contributions from

Alfred Merrill Blackmer
Gordon L. Bultena
Sharon A. Clancy
Sheila M. Cordray
Jodi Dansingburg
John W. Duffield
Derrick N. Exner
Cornelia Butler Flora
John C. Gardner
Gary A. Goreham
Chuck Hassebrook
Eric O. Hoiberg
Keith Jamtgaard

Fred Kirschenmann
Ron Kroese
Al Kurki
Harry MacCormack
Nancy Matheson
Dario Menanteau-Horta
Helene Murray
Ron Rosmann
Susanne Retka–Schill
Jim Sims
David L. Watt
Malvern Westcott
George A. Youngs, Jr.

The Initiative was an important step in profiling sustainable agriculture in the upper Midwest and Great Plains. Participants discerned dozens of important research needs to help farmers operate more sustainably and help policy-makers choose a sustainable future for agriculture and rural communities. These range from workably defining sustainable agriculture, to developing better farming practices, to improving understanding of sustainable agriculture's impact on economics, communities, and social structure. The research agenda is ambitious, but it must be measured against the importance of sustaining family farming, rural communities, and the environment.

The Northwest Area Foundation's Sustainable Agriculture Initiative was an important step in profiling sustainable agriculture and its socioeconomic impacts, rural communities, and innovative agricultural research. However, the study had limitations. A relatively small number of farmers were interviewed, generating findings that are more suggestive than definitive. Sharp distinctions between sustainable and conventional farms could not always be drawn due to variability among sustainable and conventional farmers. Considerable variation also was observed from state to state. Despite these limitations, the Initiative's research generated provocative results.

Issues raised by the Initiative transcend the four states studied, and apply to American agriculture in general. Clearly, further research of this kind is needed to explore whether patterns remain consistent over a larger sample, over a broader variety of agricultural environments, and over a larger geographic region.

RESEARCH ISSUES: DEFINING AND IDENTIFYING SUSTAINABLE AGRICULTURAL SYSTEMS

Sustainable agriculture is difficult to define. Practices that are sustainable in one bioregion may not be in another. This became quite apparent when farmers in different states were asked similar questions about their practices (Sidebar 3–3, "Which Practices Define Sustainability?"). *Consequently, research is needed to explore what constitutes sustainability in different settings. An important step is to catalog the diverse management practices used by farmers to achieve sustainability.*

One way to define sustainability is as a common set of goals across a bioregion that help sustain (indeed, regenerate) both natural and social resources through certain practices. *But we need to identify sustainable practices by further accumulating solid evidence of their environmental impact.* For example:

- What are the effects of using on-farm sources of nitrogen?
- What are the environmental implications of new crop species?
- What are the long-term impacts of intensive rotational grazing when it is accompanied by an increased concentration of livestock per acre?

Scientists, farmers, and regulators need specific measures and indicators of environmental sustainability. These must be relatively stable over time (like soil quality) and not highly variable (like annual crop yields). Examples of measures and indicators that need to be developed are:

- Above-ground indicators, such as the presence of natural enemies of pests.
- Improved measures of soil quality.
- Improved measures of energy use.
- Indicators of sustainability in livestock production.

Ideally, measures of environmental impact should be integrated to evaluate the whole farm, including on-site and off-site impacts and short-term and long-term impacts. It would be instructive to compare agricultural systems across a spectrum from regenerative to degenerative (for example, from an agriculture that rebuilds soil to an agriculture that degrades waterways and reduces biodiversity).

Whole-farm measures need to account for the environmental trade-offs of using different practices. For example, how does no-till planting affect herbicide use? How does the use of cultivation instead of herbicides affect soil quality, energy use, and the release of greenhouse gases?

How we define sustainable agriculture is critical for comparing farming systems or for predicting the impact of adoption. Clearly, one's definition affects study outcomes and policy implications (for example, see Youngs et al. 1991 and Cordray et al. 1993). This was confirmed by contrasting results from two Initiative studies:

- In Oregon, contrary to Midwestern expectations, heavy input use correlated positively with the use of the "positive practices" farmers were asked about.
- In Iowa, emphasizing reduced input use revealed a different profile of sustainable farmers than emphasizing use of positive alternative practices (compare Lasley et al. 1990 and Bultena et al. 1992).

We need to investigate further how different definitions of sustainability influence research results and policy implications.

RESEARCH ISSUES: ECONOMICS OF SUSTAINABLE FARMING PRACTICES

Farmers using apparently similar practices are achieving very different outcomes, both in income and economic efficiency. Also, sustainable farmers achieve more variable results than conventionals. Some reasons have been identified—for example, in Iowa, farmers who are members of a sustainable farm organization (Practical Farmers of Iowa) achieved better yields than nonmembers (see Sidebar 6–2, "In–Field Studies of Sustainable Farm Productivity"). In addition, two other sidebars suggest that the most successful farmers are those who employ sustainable practices most aggressively (Sidebar 6–4, "High-Return Sustainable Farms," and 6–5, "What Makes Sustainable Farms Successful? Case Studies of Three Minnesota Farms"). However, the evidence remains incomplete and other possible causes remain elusive.

To develop policies that support sustainable and moderate-scale farming, we need to learn why similar practices often produce different results, and what factors enable successful sustainable and moderate-scale farmers. *Thus, research should include in-depth interviews of farmers, with analysis of differences in their farming practices, differences in whole-farm systems, differences in management strategies and managerial skills, different uses of technology, different uses of federal commodity and conservation programs, different means of gaining information, and how more successful farmers use their labor opportunities and management skills.*

Data in the Initiative's studies were limited to a single year, subjecting the economic results to distortion from weather conditions, livestock price cycles, timing of equipment and facility purchases, and other temporal factors. *Future research should gather data over several years, which would "level" these annual variables.* This would allow identification of farms that consistently perform well and long-term factors that promote successful farming.

Data in these studies also were limited by their geographic scope and relatively small sample sizes. Although expensive, in future research it would be helpful to obtain larger and more broadly representative samples so that findings could be generalized better to the larger farm population.

This research strategy would include:

- Comparison of large and moderate-size farms that achieve high or consistently ample returns to discern whether different factors operate on farms of different size.

- Comparison of sustainable and moderate-size farms that achieve high or more stable returns with those receiving low or especially unstable returns.

These comparisons should pinpoint what makes moderate-size and sustainable farms successful, what factors weaken their economic performance, what strategies might overcome these weaknesses, which

*approaches enable moderate-scale agriculture to prosper, and what
makes sustainable agriculture economically attractive to a broader range
of farmers.* Ultimately, this comparative research would define research
and education programs and public policy to support economic opportunity and environmental stewardship.

Some less easily measured factors need examination, too. For
example:

- What is the resilience of sustainable farms to commodity-price
  fluctuation?

- What is the cost of different practices to natural resources, such
  as water quality and erosion?

- What is the potential contribution of self-produced farm resources
  (such as food and energy) to the farm's resiliency and its family's
  standard of living?

- Are sustainable or moderate-scale farms better able to provide for
  themselves, and are they more likely to do so, and how does this
  affect their chance at long-term survival?

Considering the continuing budget crisis and declining political
power of agriculture, the effect of any federal farm program change on
farm income must be projected for farms of different type and size.
Currently, the program reduces payments to farmers who grow more
diverse (nonprogram) crops on their "base" acreage. This raises important research questions:

- *How do the effects of reduced government payments to these farmers compare with the long-term economic and environmental benefits of diversification?*

- *What would happen if major policy change removed this barrier to
  more diverse production?* A redistribution of farm income would
  be one effect, but also:

- How might a significant increase in legume acreage affect other
  production elements, such as feeder cattle weight?

- What barriers (such as a "cheap grain" policy) prevent a transition to forage-based livestock operations?

*Clearly, assessment of alternative farming's environmental impact
must be integrated with assessment of its economic performance.*
Similarly, evaluation of alternative farming's economic conditions must
integrate external costs and benefits to paint a full picture. An example
would be the benefit to environment and society of sustainable farm-

ing's long–term investment in soil productivity.

However, integrated measures of socioeconomic and environmental impacts are unavailable, constituting a major barrier to assessing new technology or practices. Much more effort is needed to develop such integrated measures, and to adapt emerging methodologies from other areas of natural resource economics.

Diversification often is hailed as a "risk-reducing" management strategy, but its effectiveness requires assessment. For example, marketing constraints or opportunities can influence diversification, as noted by farmers in North Dakota and Montana. In the near term, farmers may have difficulty marketing alternative crops. *Sustainable farmers may have unmet infrastructure needs, which suggests research to determine what infrastructure they need in processing, marketing, and transportation and how it can be developed.*

Farmers may have unmet infrastructure needs because (1) they lack a well-developed market for their crops, or (2) they are trying to increase their market share by selling in "niche" markets or "differentiated product" markets. (A "differentiated product" has some unique feature that distinguishes it in the marketplace, such as a specific flavor variety grown organically.) *Where lack of market is the problem, research to develop new uses for resource-conserving crops may be essential to the success of sustainable farming.*

Consumer interest is critical to diversification, which suggests these questions for niche products:

- Are consumers aware of agricultural practices associated with the food they purchase, and do they care?

- Are consumers receptive to premium products, or to those grown on "environmentally sound" farms and sold at premium prices?

- Are consumers having difficulty with the other side of the market—finding a source for what they want? If so, how could this producer-consumer gap be bridged?

RESEARCH ISSUES: SUSTAINABLE AGRICULTURE'S IMPLI-CATIONS FOR SOCIAL STRUCTURE

The research reported here has only begun to answer these questions:

- What links exist between the practice of sustainable agriculture and rural community viability?

- How might the forces that influence rural communities be affected by a greater shift to sustainable farming?

Note that these questions *assume a relation* between farming and rural communities. More research is warranted to clarify the linkage, especially the impact of declining farm numbers:

- *Where do people go when they leave farming?*

- *What are the costs and benefits of these shifts, not only for rural communities but for urban areas?*

Among many possible linkages explored in the study, Initiative researchers found some significant differences between the two types of farming systems:

- Farms using sustainable practices tend to be smaller (except in Montana).

- Farms using sustainable practices are more labor-intensive.

- Sustainable farmers are more likely to purchase goods and services that are locally produced, and especially farmer-produced.

- Sustainable farmers spent more per acre for all locally produced or farmer-produced goods in three of the four states (in Montana, they spent much less).

Further research should clarify these relations and explore their implications.

Will the relations disclosed by the Initiative's research hold true in other states, regions, and studies? What accounts for the differences and the variability that were observed? For example, Montana's sustainable farms were equal to or larger than conventional farms studied. Is this difference an artifact of how sustainable farming was defined in the study? Or, did the difference result from the more difficult farming environment in Montana, where larger operators are in a stronger position to experiment and sacrifice their farm subsidies or to adjust their practices to a "carrying capacity" of the land that is not artificially inflated by chemical inputs?

The effectiveness, feasibility, and economics of alternatives may vary with the scale of a farm: *Is there a causal relation between sustainable practices and smaller farm size?* If so, do sustainable practices promote smaller farms, or do smaller farms promote adoption of sustainable practices? If sustainable practices were widely adopted in the Midwest, would average farm size actually decrease, or would the rate of farm size increase slow or stop?

Is it possible to identify the optimal size for sustainable farming operations with different commodity mixes and under different geographic conditions? What is the optimum? How large can operations become without sacrificing environmental sustainability?

In three states surveyed, sustainable farmers more often reported selling out-of-state, which suggests an important research topic: could alternative agriculture develop in ways that inadvertently *degrade* rural communities, for example by having to sell through large corporations that bypass the local economy? The answer depends partly on what kind of rural development policies could help develop a marketing infrastructure for sustainable farming.

How might the focus of research and technology development affect alternative agriculture's community impact? If sustainable systems are developed to be conducive to owner-operation rather than to using hired labor, and to advantage moderate-scale operations, then large-scale sustainable systems will be less likely. But, some alternative techniques may promote "bigness." For example, no-till planting reduces soil erosion, a major aspect of sustainability, but it may encourage larger farm size. *Further research is needed to evaluate and project the effects of alternative technologies on the structure of agriculture.*

Another uncertainty is the causal relation between sustainable practices and labor intensity. *The effectiveness, feasibility, or economics of alternative practices may vary over the farm family life-cycle.*

- Is the availability of family labor a basis on which farmers adjust their number of enterprises, enterprise types (crops, livestock), and farming practices? If so, how do a farm's environmental and community impacts change over the farm family's life-cycle?

- When grown children leave a sustainable family farm, does the farmer hire labor?

What are the implications of sustainable farming systems for non-family labor, especially transient labor? The study revealed little on this question, possibly because none of the study states had significant acreage in crops that require extensive hired labor for short periods. Thus, this is a research question to be addressed in other regions.

Further research and analysis is needed to better project the impact of a shift to sustainable agriculture on economic linkages. For example:

- If average farm size decreased (or remained the same), would greater population density and an increase in per-acre purchases of locally produced goods outweigh a decline of local petrochemical sales?

- What would be the impact on communities if smaller sustainable farms supported a higher population density and generated more purchases of consumer items?

Likewise, it would be useful to know how diversification at the individual farm level influences rural community viability. How do production and sales of new commodities affect local economic linkages? Does diversification pose new opportunities for rural development?

How could value-adding opportunities be increased in rural communities and especially on farms, allowing producers to more closely tie together their farm and off-farm income sources? This might require involving farmers as participants in creating new products and processes.

As each dollar injected into a local economy circulates from business to business, its economic effect multiplies. *It would be instructive to analyze trade patterns of sustainable and conventional farmers to delineate specific multiplier effects.* Current models inadequately distinguish the effects of these purchasing patterns. Thus, further research is needed to trace where money is spent and how it impacts the local economy.

Some farming communities rely on additional enterprises. Examples are timbering in western Oregon and coal mining in eastern Montana/western North Dakota. *In seeking more ways for agriculture to support such communities, it is important to know the economic share that agriculture is contributing.*

Currently, sustainable farmers are more likely than conventionals to sell and buy outside their local areas. Many farmers describe marketing bottlenecks that limit their ability to sell products, especially to add value to their products and build market linkages within their communities. Such constraints pose several research issues:

- *What is the "critical mass" of sustainable farmers necessary to develop local market linkages, such as new businesses they would buy from and sell to?*

- *How are businesses responding to new opportunities?*

- *Where are sustainable farmers successfully making local economic linkages?*

Regions seeking to link sustainable farming with sustainable development need analysis of their capacity to provide processing, marketing, and transportation resources for sustainable farmers. How can those capacities be modified by rural development policy or private initiative to meet the special needs of these producers?

Many innovative sustainable farmers also seek to improve their marketing position and increase their share with processing ventures or marketing cooperatives. *How can farmers develop joint ventures with processors to share the risk and profit?*

Viewed from another angle, how would a more widespread transition to sustainable agriculture affect the U.S. food system? If sustainable farmers significantly reduced cost and increased their share of the food dollar by selling directly to consumers, would it lead to a less centralized and less transportation-dependent food market? What is required for successful local marketing, and how can it be facilitated? Who would benefit from decentralization, and who would lose?

RESEARCH ISSUES: FARMERS' CHOICES AND CONDITIONS

Adoption Motives and Persistence. Studies under the Initiative asked farmers what year they adopted sustainable practices and their reasons for doing so (Chapter 9). However, the information collected was insufficient to clearly answer key questions:

- *To what extent are sustainable practices adopted in response to economic pressures?*
- *To what extent are sustainable practices adopted ideologically?*
- *In either case, do sustainable practices persist after adoption, regardless of economic shifts?*

Clarifying motives and persistence of adoption might help explain the large variability in economic performance among sustainable farms. Perhaps important differences exist between sustainable farmers who are ideologically committed and those who are responding to economic difficulty.

We also need a better understanding of what generates job satisfaction among sustainable farmers:

- What factors contribute to the persistence of a person in farming, both as a farmer in general and as a *sustainable* farmer?

- What expectations influence beginners to enter farming?
- Could these satisfaction factors be influenced by the design of alternative practices and technologies?

Sustainable farmers do not rely on the Extension Service as a primary source of information (Chapter 9). Instead, they rely more on their own on-farm research, other farmers, and sustainable agriculture organizations. Research of this phenomenon could help redesign Extension programs to be more useful to sustainable farmers:

- *How effectively are conventional information channels and institutions dispersing information and reaching operators of sustainable and small-to-medium-size farms?*
- *How should information channels and institutions change?*
- Is Extension failing to convey the new methods, and if so, why?
- What resources or information do Extension staff need to increase their effectiveness with sustainable farmers?
- Where extension agents or land–grant scientists are proving to be a useful source of information and sustainable farmers, what makes them successful?

Nonprofit and farmer groups appear to help farmers who wish to adopt more sustainable practices. Such organizations may contribute to how rapidly sustainable agriculture grows, so they deserve study: *What makes some organizations successful at informing and supporting farmers who seek sustainability?*

- *Could the services and support offered by farmer groups and sustainable agriculture organizations be effectively reproduced by a public organization such as Extension?*
- Can farmers who presently are disillusioned with Extension be reached by other means?

Some Initiative participants even questioned the need for an Extension Service, particularly if it continues to depend on research by land-grant universities, so part of the problem may be the current research agenda:

- *Who influences research for agriculture, and to what purpose?*
- How can universities better build coalitions with sustainable agriculture organizations and farmer groups?

SIDEBAR 15-1

Research Needs Highlighted by Study Participants

Elizabeth Ann R. Bird

Brenda J. Johnson

The research areas presented here were highlighted by Initiative participants. Most are consistent with priorities often listed in agricultural publications: crop protection, fertility management, soil maintenance, livestock production, new crops, and crop quality.

However, within these general topics, questions specifically asked by sustainable farmers, advocates, and scientists vary importantly from the agricultural mainstream. In addition, Initiative participants identified two high-priority research areas that other sources often overlook: crop/livestock integration and whole-farm management.

Here is a sampling of issues faced by farmers who try to practice sustainable agriculture.

Whole-farm Management

Farmers clearly indicated "management complexity" as an important issue with which they needed help (Chapter 5). Research to support whole-farm management must go beyond the management of specific fields or crops and *address the integration of multiple management problems.*

However, responding to management complexity is not the only issue here. Sustainable farmers are trying to replace purchased inputs and capital-intensive solutions with their own skilled labor and management. If research is targeted toward making sustainable systems more profitable, this has the potential to retain greater value, both on the farm and within the community, and to secure more farming opportunities. *Thus, to support sustainability, research is needed on management-intensive solutions to the constraints farmers face, rather than capital-intensive solutions. This may reorient research topics and how scientists frame questions.*

Crop/Livestock Integration

Sustainable farmers manage fertility and control erosion in part by integrating livestock with crops. This provides manure and makes both environmental and economic use of pasture crops and legume crops (they not only build and retain soil, but feed the livestock as well). *How can crop and livestock operations be better integrated to meet the environmental and economic goals of sustainability?*

Crop Protection

Farmers who try to reduce purchased pesticide use are seeking more information on controlling pests or weeds by manipulating farm landscape, cropping patterns, or soil-management practices. In general, farm economics and the environment will benefit to the extent that farmers are able to get nature's ecology to do their work for them. Crop-pro-

tection research is needed on:
- Strip cropping, crop diversity, crop architecture, field margins, or landscape structure to disrupt the feeding patterns of pests and to improve habitat for their natural enemies.
- Weed biology, weed ecology, and weed response to different soils, nutrients, and tillage practices.

Fertility Management

Farmers trying to reduce input cost, take advantage of natural processes, and protect the environment need more information on nitrogen and other nutrient management. Fertility-management research is needed in these areas:
- Improved use of manure and legumes as a nitrogen source.
- Better knowledge of the relation between manure and legume use and soil nitrate levels.
- Improved availability of effective soil-nitrate tests.
- Development of nitrogen tests that account for nitrogen retained in organic matter stored in the soil.

Soil Maintenance

For sustainable farmers, the issue is not merely how to contain soil erosion, but how to *improve soil quality* as well. Research is needed on:
- Strategies for maintaining the soil, such as winter cover crops that optimize the sustainability of the whole farm system.
- Means to measure, evaluate, and improve soil quality, including soil structure, soil biological activity, and effects of micronutrient levels on weed growth, health of plants, and healthfulness to consumers.

Livestock Production

New sustainable strategies are needed in dairy and livestock production to improve the farmer's bottom line while building soil, protecting water quality, and addressing social concerns:
- Refinement of management-intensive grazing strategies.
- Making grass or forage-based farming more feasible, from the beginning of livestock raising to distribution of grass-fed or forage-fed meat. Sustainability in the Midwest typically requires more forage crops in the rotations.
- Develop low–capital investment hog and poultry production methods that are profitable and address growing societal concerns about environmental degradation, consumer health, and animal welfare.

New Crops and Crop Quality

Initiative participants expressed little concern about increasing yield. They were far more concerned about developing new crops, integrating them into sustainable rotations, and improving crop quality, possibly for niche marketing. Research is needed to:
- Develop new crops that reconcile biological and economic sustainability, or improve management options for high-value crops that do so. (In western Montana, peppermint has the most lucrative gross return at $1000 per acre, but it requires intensive water and nitrogen input.)
- Evaluate new plant species for their fit into existing systems and eco-

nomic viability. (In the Corn Belt, many sustainable farmers plant oats and a legume together. However, as a cash crop, oats are not very profitable; alternatives are needed to fill that niche in crop rotations.)
• Evaluate the effects of management practices on crop quality. Such positive research findings could result in a significant marketing advantage for sustainable farmers. (Such a management practice is use of on-farm, organic resources rather than chemical inputs. Initiative researchers found that malt barley grown with green manure had a better quality processing and lower production costs--Westcott 1994).
• Determine what makes some sustainable farm systems more productive than others.
• Determine what makes some sustainable farm systems as productive, or more productive, than many of the more conventional farms.
• Notwithstanding current overproduction, develop productivity of sustainable systems to meet the *long-term* needs of a growing population.

CONCLUSION

The Northwest Area Foundation's Sustainable Agriculture Initiative has demonstrated the importance and value of farmers' involvement in designing and conducting research in cooperation with university scientists. This principle holds true for every research need identified in this chapter.

We have described a most ambitious agenda for further research. However, *this agenda should be measured in relation to the importance of sustaining family farming, rural communities, and the environment.*

16
Policy Issues and Recommendations

Chuck Hassebrook
Elizabeth Ann R. Bird
With contributions from
Brenda Johnson
Ron Kroese
Karl N. Stauber

Principal policy issues from this research include (1) farm commodity programs, (2) commodity supply management, (3) sustainable agriculture and rural development, and (4) research and extension policy. Under the latter, sub-issues are (a) addressing sustainable farmers' new information and technology needs, (b) addressing management complexity, (c) reaching beginning farmers, and (d) supporting diversity. Recommendations include (1) improving IFMPO, (2) targeting farm program payments to protect sustainable farmers, (3) creating an environmental reserve program, (4) promoting new directions for rural development policy, (5) targeting public funds to serve the public interest in sustainable agriculture, (6) providing greater support for farmer participation in research, (7) improving extension's role in supporting sustainable practices, (8) focusing research on diverse crops and new products from sustainable farms, (9) focusing research on land use by sustainable farmers, and (10) decentralizing research and extension to address local agricultural conditions.

BACKGROUND FOR POLICYMAKING

Before presenting the critical policy issues that have emerged from this research, and our recommendations, a brief recap of the Initiative's key findings is in order:

233

1. **Finding: Sustainable agriculture often is poorly served by existing public policy.** Sustainable agriculture's research needs are receiving relatively little attention in publicly funded agricultural research programs. Farmers who practice sustainable agriculture are sacrificing payments under the federal farm commodity program. (These payments are made to farmers to compensate them for low market prices for designated crops.)

2. **Finding: If sustainable farming is to succeed, its economic performance must be better than that observed in 1991.** In each state surveyed, the average sustainable farm earned lower returns than the average conventional farm. Although this reflected, in part, the effect of falling meat and milk prices that year, returns must improve if sustainable agriculture is to be more widely adopted. Wider adoption will require greater emphasis on sustainable farming in public agricultural research and extension, plus other farm policy changes.

3. **Finding: Technological, management, and policy changes that improve the economic performance of sustainable farming and result in its wider adoption are likely to increase self-employment opportunities in farming and strengthen the family farm system of agriculture.** In three of the four states, we found that sustainable farming involves smaller farms and greater use of family labor and management per acre, compared to conventional farming. If adopted throughout a community, sustainable farming would employ more families to farm the area than would be employed under conventional systems. If the sustainable farms we surveyed had rates of return on net worth comparable to conventional farms, they would have generated more net farm income per acre farmed, and per dollar invested. Thus, policies that support sustainable farming and improve its economic viability would support family farming—an often-stated objective of public policy.

   Each of these key findings encompasses important subfindings and implications. The following is a brief review of the most significant.

Sustainable Agriculture and Family Farming

- **A correlation exists between moderate-scale, owner-operated family farms and farms that use sustainable practices in three of the four states surveyed.** Family farming is characterized by *owner operation* (the owner provides most of the labor and management) and by *moderate–scale farming* (as opposed to vast acreages). The correlation with moderate–scale farming was absent in Montana, the only western state in the study—please see "The Montana Exception," below.

- **The sustainable farms surveyed were smaller on average than conventional farms in Iowa, Minnesota, and North Dakota, based on acreage, assets, and gross sales.** Furthermore, sustainable farms were more likely to remain smaller, because they were less likely to plan expansion. Despite the smaller size of their operations, the sustainable (SUST) farmers surveyed had farming as their principle occupation just as often as conventional (CONV) farmers. Sustainable farm operators were no more likely to work off-farm than conventionals.

- **Greater labor requirements limit the size of sustainable farms.** Sustainable farms in Iowa, Minnesota, and North Dakota used more labor per acre than conventional farms, even adjusting for the number of livestock enterprises on the farm. (But in Montana, this was not the case—see "The Montana Exception," below.)

- **Despite greater labor requirements, sustainable farms in three of the four states relied less on hired labor than conventional farms.** This was measured as the percentage of labor provided by non-owner employees from outside the household. The greater labor need usually is met in a manner consistent with family farming: by moderating farm size, rather than by moving toward an industrial structure that relies on non-owner employees for labor. This may reflect the more balanced labor demand throughout the year that is characteristic of sustainable farms, making family provision of labor more feasible.

- **Sustainable farmers cited crop and soil management as their greatest challenge in acquiring new knowledge, whereas conventionals cited marketing.** The demand for careful crop and soil management may encourage the operator to spend more time in the field managing operations, rather than relying on less-motivated, less-knowledgeable employees to do so. By contrast, the marketing challenge cited by conventional farmers is an area more easily isolated from farm operation. This allows the conventional farm owner to focus management on marketing, leaving less management-intensive farm operations to employees.

235

- **The greater labor and management requirements of sustainable farms can reinvigorate moderate-scale family farming—if farmers receive an adequate return on the additional labor used.** However, the average sustainable farm surveyed generally did not receive an adequate return on labor in 1991. Iowa was the only state where return on net worth among SUST farmers was within 4 percent of CONV farmers. There, SUST farmers captured a greater share of gross farm income as net farm income, and they earned more than double the net income/acre of conventional farms. Yet, as a group, SUST farmers surveyed in all four states were unable to pay $5 per hour for the additional labor involved and still show a return on net worth comparable to conventional farms.

  Nonetheless, the finding suggests that, if public policy changes were to make sustainable farming more economically viable, the results likely would help reinvigorate family farming. (Examples of such change would be to reduce the bias against SUST farmers in federal farm commodity programs, and to correct the imbalance in the division of resources between conventional and sustainable agriculture in agricultural research and extension programs.)

- **Increased adoption of sustainable systems would probably slow the growth of farm size, due to the greater management and labor requirements, and more family farming opportunities likely would occur.** As in Iowa, farmers would capture a greater share of agricultural dollars as net income, and those dollars would reinvigorate farm communities.

- **More young people might be attracted to farming if they could use their skills, labor, and management to replace capital and retain a larger share of gross income as net income.** Capital typically is the most limiting resource for beginning farmers, so the opportunity to replace capital with skilled labor and management in many cases fits their resources and limitations.

The Montana Exception

As noted, the research in Iowa, Minnesota, and North Dakota correlated sustainable practices with moderate–size farms and greater labor use, but the Montana research revealed the opposite. In Montana, the sustainable farms sampled were larger, planned more expansion, and used less labor per acre farmed (in the case of ranches and irrigated farms) than conventional farms.

The Montana findings were complicated by the state's diverse agriculture, ranging from smaller irrigated farms in river valleys, to dryland grain farms, to extensive livestock ranches. For example, labor requirements per acre for ranches and irrigated farms were *less* for sustainables than for conventionals. An explanation for irrigated farms may be that more conventionals specialize in high-labor crops, particularly hay.

Montana's sustainable dryland farms were the opposite; their farm labor/acre requirements were about 50 percent *more* than for conventionals. Montana sustainable farms were consistent with the family farm concept, because they relied on outside employees for a smaller percentage of their labor than conventional farms.

Clearly, more research is needed to understand the implications of sustainable agriculture in western states.

FARM AND RURAL DEVELOPMENT POLICY ISSUES

We believe the following policy issues, disclosed by the research findings, are critical to sustaining family farms, rural communities, and the environment. We accompany each issue with one or more recommendations.

POLICY ISSUE 1

Farm Commodity Programs

The research discovered a clear bias against sustainable farms in federal farm commodity programs. This bias is most evident in the crops that qualify for deficiency payments. Under the farm commodity program, crops that qualify ("program crops") are primarily corn, grain sorghum, wheat, barley, cotton, oats, and rice. "Deficiency payments" are made by the U.S. Department of Agriculture (USDA) to farmers as compensation for shortfalls between market prices and congressionally established "target prices" on selected crops.

For example, in the survey year (1991), the target price for wheat was $4.00 per bushel, but the actual market price was only $2.70 per bushel. Thus, farmers growing wheat had a shortfall of $1.30 per bushel, qualifying them for this amount in deficiency payment on a portion of their production.

The problem for SUST farmers is that many crops they typically grow are excluded from the legislation that establishes deficiency payments. Examples of excluded crops are forage plants such as alfalfa and clover; small-acreage crops such as rye, millet, a rye-wheat hybrid called triticale, and amaranth; and even soybeans. Furthermore, the target price has been set so low for oats, another crop often grown by sustainable farmers, that significant deficiency payments are rarely made.

In each of the states surveyed, sustainable farms had a smaller percentage of acreage than conventionals in crops that qualify for deficiency payments:

- In Iowa, farm acreage that was planted in crops qualifying for deficiency payments was 64 percent for conventional farms but only 45 percent for sustainable farms.

- Acreage data for Minnesota were unavailable, but commodity program payments to CONV farmers there were nearly three times those made to SUST farmers, measured as percent of gross sales. We infer that CONV farmers plant substantially more acreage in qualifying crops than sustainables.

- In Montana, CONV farmers planted qualifying crops on 75 percent of their acreage, versus 62 percent for SUST farmers.

- In North Dakota, qualifying crops were planted on 67 percent of conventional farm acreage but only 49 percent of sustainable acreage.

Oats were not counted as a qualifying crop in this analysis, even though technically oats are eligible, because significant deficiency payments rarely are made. This bias in deficiency payments has important negative consequences for sustainable agriculture:

- The deficiency payment bias reduces sustainable farm profitability and discourages sustainable practices.

- The deficiency payment bias disadvantages SUST farmers in surviving the economic roller coaster of agriculture.

- The deficiency payment bias disadvantages SUST farmers in competing with CONV farmers for land and markets for their products. All other things being equal, the inequity in deficiency payments gives a SUST farmer less money than a CONV farmer to pay for land. Consequently, many SUST farmers are bid out of the land market.

- The deficiency payment bias discourages beginning farmers from using sustainable agriculture as an entry strategy.

- The deficiency payment bias discourages CONV farmers from converting to sustainable farming for fear of losing payments.

The long-run effect of this bias is to drive some SUST farmers out of farming who otherwise would not leave, to discourage other farmers from adopting sustainable practices, and to make it harder for beginning farmers to use low-capital sustainable strategies.

Recommendation 1A: Improve IFMPO

Several 1990 Farm Bill provisions were passed to give farmers greater flexibility in converting to sustainable crop rotations. Most notable is the *Integrated Farm Management Program Option (IFMPO)*. Under IFMPO, certain resource-conserving crops can be planted on acres normally reserved for program crops, but without losing deficiency payments. However, the IFMPO program has been plagued with administrative problems and overly restrictive legislation. It has had limited effectiveness in removing barriers to sustainable farming. (IFMPO is described further in Sidebar 16-1, "Alternative Policy Options").

SIDEBAR 16-1

Alternative Policy Options

Thomas L. Dobbs

Agricultural economists at South Dakota State University (SDSU) and Washington State University (WSU) analyzed these policy options to encourage more sustainable practices in the northwest region:
1. A tax on purchased chemical fertilizers and pesticides.
2. Mandatory supply controls through stringent restrictions on planted acreage.
3. Various policy options to increase the planting flexibility for farms.

In South Dakota, researchers used economic models of paired case farms to analyze policy impacts. In the Palouse region of eastern Washington State and northern Idaho, economic models of farms based on synthesized data were employed. Analytical methods differed somewhat between the South Dakota and Palouse studies, and not all policies were analyzed in both regions.

Together, however, the SDSU and WSU studies provide valuable insight into the possible future direction of U.S. farm policy, and implications for sustainable farming in the northwestern region. These insights are summarized briefly here.

25 Percent Tax on Fertilizer and Pesticides

In South Dakota, SDSU postulated a tax (25 percent of retail price) on commercial fertilizer and pesticides and analyzed its impact on farm profitability. The effect on profitability was greatest on conventional farms in eastern South Dakota, where corn and soybeans dominate and where soil and moisture conditions are more conducive to intensive chemical use.

In general, however, a 25 percent tax did not appear sufficient to

---

make farmers switch from conventional to sustainable farming—*except* where the systems were of nearly equal profitability to begin with, which sometimes appears to be the case in wheat-growing areas of South Dakota. Of course, such a tax could induce conventional farms to *reduce* their fertilizer and herbicide application rates *without* adopting other sustainable practices or completely changing their crop rotations.

Policy: Mandatory Supply Control

Mandatory supply control, implemented through severe restrictions on planted acreage of "program" crops (including soybeans), was analyzed using case-study farms in South Dakota. This policy favored conventional farms, primarily because of the very high prices induced by the restrictions on crops that dominate conventional farming (corn, soybeans, and wheat).

Mandatory acreage controls tend to increase the application of commercial chemical inputs to boost yield from the remaining acres to benefit more fully from higher prices, but a tax on these inputs might partially offset that effect. In principle, one could design a mandatory acreage-control program that also requires sustainable practices like the inclusion of legumes in crop rotations.

Policy: Planting Flexibility

Normal Crop Acreage (NCA). Greater flexibility in choosing crops to plant would enable farmers to adopt more sustainable rotations. Proposals to increase planting flexibility were discussed for the 1990 Farm Bill. Although not adopted, the Bush Administration proposed a Normal Crop Acreage (NCA) program.

Normal crop acreage for a farm is established by summing individual crop acreage bases and historical plantings of oilseed crops (soybeans, sunflower, rapeseed, canola). Any combination of program crops and oilseeds may be planted on the normal crop acreage without losing present or future program payments. Government deficiency payments under this program would be based on historical plantings and base yields, except for deductions based on harvested acres of nonprogram or nonoilseed crops on NCA. In other words, government payments would be largely "decoupled" from what farmers actually grow.

SDSU researchers analyzed a second version of the NCA program, in which harvesting of legumes and other nonprogram crops (such as millet and buckwheat) planted on the NCA base would be permitted without reducing deficiency payments. In both versions, legumes or other crops could not be harvested from set-aside acres.

Results indicate that NCA proposals encourage diversified cropping. However, where conventional corn and soybean production is relatively profitable, as in parts of eastern South Dakota, the NCA options by themselves appear insufficient to induce conversion from conventional to sustainable cropping systems.

In wheat-growing areas of northern and western South Dakota, however, conventional and sustainable farming sometimes are of nearly equal profitability. Here, NCA policies could significantly influence conversion from conventional to sustainable farming, particularly if deficiency payments were not reduced for harvesting legumes and other nonprogram crops on the NCA base.

To achieve this positive effect on sustainable farming profitability, NCA policies may need to be introduced gradually to limit adverse market effects for legumes and other nonprogram crops which are important in the rotations of existing sustainable farmers.

WSU research concluded that an important NCA program benefit would be largely to eliminate the decline over time of farm-program "base acreage," which now occurs when farmers adopt green-manure rotations. (At present, a farmer participating in the standard program can gradually lose "base" acres of a crop, such as wheat, by involving extensive green-manure acreage in the rotation.) This would relax one constraint to wider use of sustainable practices.

IFMPO. In the final 1990 Farm Bill, a complex pilot program was approved: the Integrated Farm Management Program Option. IFMPO, a voluntary commodity program, gives farmers greater flexibility to develop more diverse, resource-conserving crop rotations. To participate, a farmer must plant at least 20% of crop acreage base in resource-conserving crops. IFMPO provides farm-program payments for planting resource-conserving crops on acres that are eligible for deficiency payments. It also allows some harvesting of set-aside acres.

As WSU research disclosed, IFMPO has slightly strengthened farmer incentives to use sustainable cropping systems, and results in a South Dakota wheat-growing region were similar. However, in the corn-soybean areas of South Dakota, IFMPO does not begin to offer adequate economic incentive for farmers to make major shifts to sustainable farming. IFMPO can facilitate some incorporation of resource-conserving crops in corn-soybean systems.

Other Flexibility Options. Six other flexibility options were analyzed for the Palouse region by WSU researchers:
1. The 1990 Farm Bill.
2. The 1990 Farm Bill, with unpaid flex acres increased to 40%.
3. The Bush Administration's NCA proposal.
4. *Decoupling*—complete separation of farm program payments from crop-planting decisions.
5. *Recoupling*—separation of farm program payments from crop-planting decisions, but graduated program payments would be recoupled to meeting certain criteria for conservation and environment. This recoupling would be a form of "stewardship" or "green" payments.
6. A free-market scenario, in which current commodity programs would end.

These policy options were compared from four standpoints—farmers' profits, consumer benefit in lower food prices, cost to taxpayers, and environmental damage. Net social welfare associated with each policy was measured by combining monetary estimates of costs and benefits for all four categories.

When crop prices are average, WSU research found little variation in net social welfare among the policy options. However, the *distribution* of benefits and costs among farmers and other affected groups varied considerably. For example, estimated net returns to farm operators were lowest under the free-market scenario and highest under the NCA program, assuming conventional cropping systems. Environmental damage was reduced most under the decoupling, recoupling, and free-market scenarios when alternative cropping systems were made available.

High grain prices could alter this conclusion, however. In this scenario, only the *recoupling* policy option succeeded in protecting the environment, because only the recoupling option would explicitly pay farmers to reduce soil erosion and nitrogen use. Recoupling ranked highest in estimated net social welfare in the Palouse when researchers assumed high grain prices.

In other words, when high grain prices draw many farmers out of commodity programs, the most cost-effective policy to achieve given environmental benefits may be to pay farmers on the basis of their stewardship.

FOR FURTHER INFORMATION

Details of the research methods and findings are in Dobbs (1992); Dobbs and Becker (1992); Dobbs, Taylor, and Smolik (1992); Painter (1992); and Painter and Young (1993). WSU researchers also analyzed impacts of policy options in the North Carolina Coastal Plain; those results are not described here but are available in Painter and Young (1993).

IFMPO could be more effective if restrictions were loosened on harvesting resource-conserving forage crops planted by IFMPO participants on acres normally planted in program crops. In the three states for which data are available, SUST farmers devoted a larger percentage of their cropland to resource-conserving forage crops than did conventionals: North Dakota, 3 percent (versus conventionals, less than 1 percent); Iowa—10 percent (conventionals, less than 1 percent); Montana—19 percent (conventionals, only 4 percent).

In Minnesota, more than twice as many SUST farmers raised alfalfa as CONV farmers, committing about four times the acreage (as a percent of acres operated). Half the Minnesota SUST farmers raised other types of hay, whereas only one of the nineteen CONV farmers did.(The Minnesota survey distinguished between alfalfa hay acreage and "other hay" acreage; the other states did not.)

Recommendation 1B: Target Farm Program Payments to Protect Sustainable Farmers

Changes to IFMPO cannot alleviate the competitive disadvantage suffered by long-time SUST farmers, because they have *smaller historical bases* of program crop production. Like farmers in any commodity program, farmers who adopt IFMPO receive payments for base acres only (their acreage historically planted in program crops).

A fair way to alleviate this competitive disadvantage is to protect them from reductions being made in farm program benefits to reduce federal deficits. SUST farmers already have suffered a cut in commodity program benefits by farming sustainably and thereby sacrificing base acres and deficiency payments. Approaches for cutting farm program spending while simultaneously protecting sustainable farmers include:

- Cut program cost by reducing the maximum payment that one farmer can receive (currently $100,000).
- Cap the percentage of a farmer's crop acres on which payments can be received, perhaps at 70 percent of crop acres.

Either would substantially cut farm program cost by reducing payments to the biggest recipients of farm program benefits, and at the same time insulate from additional cuts those SUST farmers who receive meager payments. These measures would reduce the discrepancy in benefits between sustainable and conventional farms. SUST farmers would face less policy-induced disadvantage in competing for land, survival, and even entry into agriculture.

POLICY ISSUE 2

Commodity Supply Management

To receive deficiency payments, farmers typically idle part of their acreage, which limits commodity supplies and thus strengthens market prices. But this approach reduces land use and, in response to higher prices, also encourages greater and more intensive production on the remaining active acres, with concomitant increases in use of yield-increasing chemical inputs. Therefore, this system is biased against SUST farmers, because they typically use land less intensively than CONV farmers, use fewer purchased inputs, employ more crop rotation, and plant fewer acres in program crops.

Ultimately, SUST farmers are forced to make multiple contributions to supply control. They must idle land to comply with commodity program rules if they wish to receive deficiency payments, but at the same time they contribute to supply-management objectives by using practices that reduce program-crop acreage and, in some instances, reduce yields.

In Iowa, Montana, and North Dakota (but not Minnesota), SUST farmers had somewhat lower 1991 program crop yields than CONV farmers. This makes SUST farmers appear inefficient—but lower yield and inefficiency are not synonyms. Given the federal objective of managing the supply by reducing production while simultaneously protecting the environment, sustainable farming actually may be *more* efficient than applying chemical inputs to achieve maximum yield on some acres while idling others.

Recommendation 2: Create an Environmental Reserve Program

A strategy that would be fairer to SUST farmers, that would benefit the environment, and that could be more efficient is a shift to an *environmental reserve*. Through this, farmers would be paid to reduce production, but *in ways that benefit the environment*. Eligible activities could include restoring wetlands, creating contour grass strips for erosion control, adding soil-building crops to rotations, installing filter-strips of grass along waterways, and setting lower yield goals with concomitant reduction of nitrogen and pesticide use.

Farmers would voluntarily bid to reduce production, with USDA accepting bids offering the most environmental benefit and supply control per dollar spent. Multiyear enrollments (three to five years) would maximize environmental benefit and justify the cost of establishing practices. Limited one-year enrollments also could be accepted.

Some current deficiency payment funds would be shifted to this program. Eliminated would be the requirement that all commodity program participants idle land as a condition of receiving deficiency payments.

POLICY ISSUE 3

Sustainable Agriculture and Rural Development

Widespread adoption of sustainable agriculture would change farm community economies and present different opportunities. The data suggest that two trends would result: (a) local retail sales of *non-locally produced inputs* would suffer, particularly farm chemicals, and (b) sales of *locally produced farm inputs* would remain as they are today or increase.

All else being equal, sales of locally produced inputs contribute more to a local economy than inputs manufactured elsewhere. For locally produced inputs, value is added locally, and a larger share of the sale price is retained in the community. In contrast, for inputs manufactured elsewhere, all that is retained locally is the retail markup.

Likewise, when inputs can be reduced without lowering crop yields, more farm income is retained, which benefits local communities by making available more money for consumer purchases and investment. (For more on retention of farm dollars within local communities, see Chapter 7.)

Widespread adoption of sustainable practices likely would slow declining farm numbers. Simply reducing farm consolidation would boost social and economic viability of agricultural communities (Swanson 1980). Of course, sustainable farming's adoption depends on demonstration of better economic performance compared to the farms studied in 1991.

Recommendation 3: Promote New Directions for Rural Development Policy

Inputs. SUST farmers use similar or greater amounts of locally produced and farm-produced inputs per acre, compared to *CONV* farms. Thus, rural development policies should facilitate economic linkages, both between farmers and local businesses that produce inputs, and among farmers themselves. Such linkages could help communities overcome the loss of retail farm chemical sales associated with conversion to sustainable agriculture.

Marketing. Nearly half of the SUST farmers surveyed in Montana and North Dakota expressed strong concern about markets after they had gained experience with sustainable farming. In fact, SUST farmers were more likely to sell their goods out-of-state for lack of local markets.

This suggests niche market opportunities, possibly for organic production and for small-acreage crops with limited markets. The small size of these niches reduces their attractiveness to large regional processors located in population centers, creating small business opportunities in farm communities.

Although small, these opportunities could represent a number of good-quality economic opportunities. Today, businesses that produce high-value specialty products which are unavailable to mass markets are earning higher returns than mass producers of "undifferentiated" products (Ikerd 1992). This creates possibilities for sustainable agriculture in lean meats, natural meats, organic crops, and unique food crops such as blue corn for chips. Production of each of these products fits well in sustainable systems.

To fully realize high-return opportunities for these specialty products, support is needed from state and federal programs in economic development and agricultural marketing, cooperative research, and extension. Support must focus on moderate-scale applications to enhance opportunities for family farms and small businesses in farm communities.

Of course, SUST farmers will continue to be involved in producing undifferentiated commodities. The finding that sustainable farmers are smaller and have more small enterprises suggests that they face special marketing problems, because the commodity market pays premiums to larger-volume producers and less to smaller producers. Public policy must provide these smaller-volume producers with means to overcome this disadvantage. Examples are:

- Legal remedy, such as antitrust rules against discriminatory pricing.

- Support for new cooperatives aimed at increasing the market power of SUST farmers and other small producers.

RESEARCH AND EXTENSION POLICY ISSUES

The sharp imbalance in public agricultural research and extension programs must be corrected if we are to improve the economic performance of sustainable systems and widely realize their social and environmental benefits. Currently, sustainable systems receive relatively little support.

USDA's primary research program in this area is the **Sustainable Agriculture Research and Education Program (SARE)**. SARE currently receives only *one half of one percent* of the annual federal expenditure on agricultural research and extension. Further, sustainable agriculture has received quite limited attention in mainline USDA research programs and in state university budgets (Bird 1991; Bird and Hassebrook 1992; National Research Council 1989; Lockeretz and Anderson 1993). Recent self-analysis by USDA's in-house research group, the Agricultural Research Service (ARS), indicated that only 21 out of 1,177 projects fully qualified as "sustainable agriculture" research. These accounted for about one percent of the ARS budget (Bird 1994).

This imbalance is critical, because the direction in which research and education dollars are invested determines which farming system will enjoy full development, refinement, economic competitiveness, and eventual adoption by farmers. If sustainable agriculture is to strengthen economically and attain wider adoption, and if we are to realize its social and environmental benefits, sustainable agriculture must have stronger support from public research and extension programs.

In fact, our findings indicate that such increased support would fulfill two major policy directives of the 1990 Farm Bill, formally titled *Food, Agriculture, Conservation and Trade Act of 1990 (FACTA)*. Here is a summary.

1. FACTA's section on *Research and Extension Purposes* directs USDA to emphasize research aimed at:

    • Increasing economic opportunities in rural communities.

    • Improving quality of life for farmers and society as a whole.

    • Enhancing the environment and natural resource base.

    • Enhancing human health.

    • Promoting more traditional goals, such as improving competitiveness and productivity.

    (*Congressional Record*, October 22, 1990)

2. FACTA directs all competitive federal grant programs to emphasize research consistent with sustainable agriculture, defining it as:

    . . . integrated production systems that meet food and fiber needs, enhance environmental quality and the natural resource base, make efficient use of on-farm resources and nonrenewable resources, integrate biological cycles and controls, are economically viable for farmers, and enhance quality of life.

    The concluding phrase of that definition, calling for *enhanced quality of life*, was further defined in a statement by Agriculture Committee leadership, calling in part for:

    . . . research which increases income and employment—especially self–employment—opportunities in agriculture and rural communities and strengthens the family farm system of agriculture, a system characterized by small and moderate-sized farms which are principally owner-operated.

    (*Congressional Record*, August 1, 1990)

    Research to improve sustainable agriculture's economic performance is likely to increase opportunities in family farming and farm communities. *Research programs that strengthen sustainable farming would help meet these Congressional directives.*

POLICY ISSUE 4

Research and Extension: Addressing Sustainable Farmers' Need for New Information and Technology

Sustainable farmers rely less on purchased inputs and more on management. This implies a new research need, to create *knowledge* that SUST farmers can use to manage more effectively and reduce their need for purchased inputs. Unfortunately, such research is unattractive to private industry because it does not translate into product sales.

Recommendation 4: Target Public Funds to Serve the Public Interest in Sustainable Agriculture

Sustainable agriculture requires "public-good" information and technologies. Public-good technologies are widely available and benefit the broad public, in contrast to those that offer profit opportunity for their developers, such as input-supplying agribusinesses or patent-holding public universities.

Examples of such technologies include information to better manage habitat for beneficial insects, improved composting approaches or rotation strategies that eliminate the need for chemical fertilizers, and crop seed varieties that farmers can reproduce themselves. Public funds should be targeted toward developing public-good information and technologies.

Generating the Knowledge. Providing the knowledge and technologies needed for sustainable systems will require ample public investment in both fundamental and applied research. Fundamental research examines how and why natural phenomena work as they do, whereas applied research attempts to solve concrete problems, often using the understanding gained through fundamental research.

An important trend has been to conduct fundamental research that supports private-sector technology development, and to leave much applied research to the private sector. If public research institutions are to support sustainable agriculture, this trend must be reversed. Both fundamental and applied research should be directed toward developing public-good technologies that improve sustainable systems.

Integrating Research. Research traditionally is designed to occur in a sequence: fundamental research, then developmental and adaptive (applied) research, followed by technology transfer. But with this approach, it is hard to be sure of where the effort is going until we get there. To ensure that the end result is a sustainable agriculture, all research phases should be integrated. One way to integrate fundamental and applied research is to conduct them simultaneously. The researchers involved in these parallel lines of research should constantly communicate and collaborate.

To achieve this integration at the federal level, research could be much better coordinated between the USDA's National Research Initiative (NRI), a competitive grant program mostly for fundamental research, and the USDA's Sustainable Agriculture Research and Education Program (SARE), a competitive grant program mostly for applied and on-farm research.

The 1990 Farm Bill directed NRI to emphasize sustainable agriculture and apply at least 20% of its funds to "mission-linked systems research." SARE could help identify priorities for NRI research in sustainable agriculture. NRI could use the SARE regional evaluation structure of its mission-linked systems research, and the two programs could jointly fund projects.

Another option is to require proposals and reports for fundamental research to demonstrate tangible links to applied researchers and farmers, and to specify potential contributions to solving real farm problems that are important to sustainable agriculture.

POLICY ISSUE 5

Research and Extension: Addressing Management Complexity

The SUST farmers interviewed said that sustainable agriculture involves more management and complex decision making than conventional agriculture. Management skills were a strong concern among SUST farmers in Iowa and Minnesota. Likewise, weed management was the top concern in three of the four states and the second-ranking concern in the fourth state. The availability of information also was a top concern.

If sustainable agriculture is to be widely adopted, farmers will need more help from research and extension programs in developing the knowledge and skills to manage their farms successfully. Yet, when SUST farmers ranked their sources of information, conventional information sources such as land-grant research and extension institutions fared poorly. In all four states, farmers said they gained more from their own on-farm research, other farmers who practice sustainable agriculture, and sustainable farming organizations.

Overall, personal on-farm research received the highest usefulness rating among SUST farmers. Between 75 and 90 percent of farmers in Iowa, Minnesota, and North Dakota ranked their on-farm research as very useful; 64 percent of those in Montana did so.

Sustainable farmers generally rate on-farm research and nonprofit sustainable agriculture organizations as more useful sources of information than public research and extension. This finding suggests two possibilities:

1. Both research and extension need to address the complexities of whole-farm management, and better attend to the farmer's role in shaping the farm system.

2. Extension might improve its effectiveness by working more closely with the sustainable agriculture organizations that farmers rated as more useful and by adopting their approaches. These approaches include participatory research, support groups, collective problem-solving, and information-sharing. Extension also should consider contracting with these groups to provide services to farmers (see Chapters 9 and 10).

Recommendation 5A: Provide Greater Support for Farmer Participation in Research

Scientific research and the practical on-farm research of farmers must be integrated to address the managerial complexity and site-specific nature of sustainable farming. Full participation of farmers in research is critical to achieving this integration. Such participatory on-farm research must focus on problem-solving and improving the whole-farm system.

To realize sustainable agriculture's potential, public research and extension programs must support this participatory research. Chapter 10 describes several innovative models for participatory research.

Unfortunately, participatory and interdisciplinary whole-farm research often go unrecognized and unrewarded by traditional academic promotions, acceptance of journal papers, and acknowledgment by professional societies. These institutions must be reformed so that academic researchers are not penalized for their efforts with SUST farmers and their organizations.

Recommendation 5B: Help Farmers Develop Management and Decision-Making Skills

Helping farmers to develop management and decision-making skills involves more than providing information and transferring technology. It requires working actively with farmers to help them develop critical decision-making skills and consulting with them on individual management dilemmas. An emerging technique is the decision case, described in Chapter 11.

For research and extension programs to help increase family farm opportunities, they must reach beyond the most aggressive and capable farmers to those who need more skill development. Accomplishing this requires innovative programs, at a time when extension funding is being reduced in many states. But if farmers are to acquire the skills they need, a strong, publicly funded extension system must be maintained.

POLICY ISSUE 6

Research and Extension: Reaching Beginning Farmers

In each state, between 40 and 50 percent of SUST farmers adopted sustainable practices before age thirty. Around 80 percent adopted sustainable practices prior to age forty. Clearly, most farmers who use sustainable practices adopted them while younger. This suggests that young and beginning farmers are a critical target for sustainable agriculture extension.

Recommendation 6: Target Beginning Farmers

To reach the farmers most open to sustainable agriculture, extension must target beginning farmers. Extension programs should be designed to fit the needs and limitations of these beginning farmers. Research may be needed to identify the best ways to reach them.

To help increase farming opportunities, research should emphasize low–capital investment alternatives, such as pasture–based livestock enterprises.

POLICY ISSUE 7

Research and Extension: Supporting Diversity

A majority of CONV farmers in the four study states raised one or two crops, but SUST farmers raised three or more. Thus, SUST farmers have more diverse enterprises. SUST farmers also were more likely to experiment with nontraditional crops, such as peas, spelt (a wheat used as livestock feed), and amaranth. SUST farmers also were much more likely to plan expansion of the number of crops being grown, and to include livestock.

The diversity among sustainable farms and their operators is amply demonstrated in the Initiative's survey as well. Sustainable farming systems within bioregions often share common practices, but each sustainable farm is managed as a whole agro-ecosystem and consequently is unique. To support this diversity, which so characterizes sustainable agriculture, we offer several recommendations.

Recommendation 7A: Focus Research on Diverse Crops, Rotations, and Genetics

Today, agricultural research concentrates heavily on conventional large-acreage crops such as corn, wheat, and soybeans. But the survey's findings suggest that NRI and other research programs that support genetic research should focus on making improvements to more diverse crops, especially those grown in rotations used by sustainable farmers. This would support NRI's congressionally mandated "sustainable agriculture emphasis."

Recommendation 7B: Focus Research on Developing New Products from Sustainable Farm Crops

Sustainable agriculture needs research to develop new markets for products from resource-conserving crops and rotation crops, including small grains and forage legumes.

Recommendation 7C: Focus Research on Land Use by Sustainable Farmers

For SUST farmers, it is important to find ways to use diverse landscapes that are both sustainable and economic. For example, research is needed on sustainable/economic ways to use woods, wetlands, and grasslands.

Recommendation 7D: Decentralize Research and Extension to Address Local Agricultural Conditions

Sustainability is achieved differently on each farm. Because sustainable farms are so individual, site-specific, diverse, and evolving, public research and extension need to address the unique conditions of different areas and farms. Because a farmer's knowledge of his or her farm is so important to sustainable management, researchers and extension agents should work closely with farmers to integrate local and scientific knowledge (see recommendation for participatory research, above).

The present centralized control over local research and extension is a barrier to the site-specific research needs of sustainable agriculture, according to university researchers who participated in the Initiative. Researchers working in off-campus experiment stations argued for more freedom to pursue locally defined problems. Off-campus researchers also felt that collaborative efforts with on-campus colleagues were strained by their difference in orientation.

The absence of funds presents another barrier to participatory, site-specific research. State support for agricultural research and extension should ensure that funds are available to develop sustainable systems in each bioregion.

RESEARCH AND EXTENSION POLICY IN PERSPECTIVE

Each recommendation provided above would improve the environment and opportunity for sustainable agriculture research and extension. The central tasks are:

- To better direct public research and extension funds toward support for sustainable agriculture research.
- To enable innovative research approaches and cross-functional collaborations.
- To develop institutional frameworks that fully involve farmers, nonprofit organizations, and others in setting the research and extension agenda, and implementing it.

CONCLUSION

The research suggests that sustainable farming can enhance environmental quality and economic opportunities in family farming—*if policy is reformed to create a level playing field and strengthen the economic performance of sustainable farming.*

APPENDIX A
STUDY METHODS

Keith Jamtgaard

Note: The study methods are summarized in Chapter 3. This appendix provides a more detailed explanation.

Researchers in each of the four states engaged in two data-gathering phases:

- *Phase I*—telephone interviews and mail questionnaires.

- *Phase II*—in-depth interviews and mail questionnaires administered to smaller subsamples of Phase I farmers. (Researchers from Oregon State University also conducted telephone surveys of producers in western Oregon and Washington as part of this initiative, described in Cordray and Goetz 1991, but did not participate in Phase II.)

Because our Phase II research samples were drawn from Phase I, we describe the sampling procedures for each phase below.

Phase I: Questionnaire Survey

All four states used similar procedures to develop their samples for Phase I. Farmers were selected in two steps. First, a "basic farm sample" was drawn to represent all farmers in the selected areas (Table A-1). The second step drew a "supplementary farm sample" (Table A-1) of farmers interested in, or thought soon to be adopting, sustainable agricultural practices.

In contrast to most previous research, organizational membership was not used in classifying a farm as sustainable. Rather, this determination was based on farmers' scores on a *sustainability index.* Many farmers who ranked high on this index came from the supplementary samples. Sampling procedures used in each state accompany Table A-1. Response rates for the surveys are presented in Table A-2.

Table A-1. Procedures used to develop Phase I samples.

Parameter	Iowa	Minnesota	Montana	North Dakota
Sampling description	Two-stage— 3 counties selected within each of 5 regions; randomly selected farmers	Single-stage (farmers)	Single-stage (farmers)	Single-stage (farmers and ranchers) (Leistritz and others 1989)
Basic farm sample	Farm & Home Directory	Minnesota Agricultural Statistics Service list	Montana Agricultural Statistics Service list	Leistritz and others (1989)
Supplementary farm sample	Membership lists from 4 sustainable farm organizations	Additional sampling from 9 counties where educational programs are underway; mailing lists from 3 organizations	Mailing list from alternate Energy Resources Organization (AERO)	Membership list from 1 organic organization
Screening criteria	50 acres or more; raised corn in 1989	50 acres or more	50 acres or more	Under age 65; farming considered primary occupation; sold $2,500 minimum farm products in 1984

NOTES

Iowa. Two-stage sampling identified both farm and supplemental samples (details: Bultena et al. 1992). In the first stage, each Iowa county (N = 99) was assigned to one of five agricultural regions. Three counties were selected randomly from each region, yielding 15 study counties. For the farm sample, a second stage randomly drew respondents from each county, proportional to the number of farms in that county but excluding operating farms < 50 acres. Respondents were telephone-interviewed and completed mail questionnaires. A supplemental sample was obtained from membership lists of four Iowa sustainable farming organizations, excluding operating farms < 50 acres and including only those that raised corn during 1989.

Minnesota. Single-stage sampling identified the farm sample (Menanteau-Horta and others 1991). The Minnesota Agricultural Statistics Service (MASS) randomly invited participation from farms statewide. The supplemental sample was obtained from several sources. MASS developed an additional sample from nine Minnesota counties that had educational programs to increase sustainable agriculture awareness, on the expectation that more farmers in these counties would have been exposed to sustainable agricultural ideas. Other sources were membership or mailing lists from two institutions working with sustainable agriculture in Minnesota and the Minnesota addresses from a sustainable agriculture magazine's mailing list. Questionnaires were screened to exclude operating farms smaller than 50 acres.

Montana. Single-stage sampling identified the farm sample (Jamtgaard 1992a). The Montana Agricultural Statistics Service randomly invited farmers and ranchers having operations larger than 50 acres to participate. A supplementary sample was obtained from the Alternative Energy Resources Organization (AERO), which maintains a mailing list of Montanans who have expressed interest in sustainable agriculture.

North Dakota. Single-stage sampling identified the farm sample by random selection of North Dakota farmers and ranchers involved in an earlier study (Watt and others 1992). During the earlier study, a random sample of farmers and ranchers had been screened to "ensure that all respondents were less than 65 years old, operating a farm, defining farming as their primary occupation, and sold at least $2,500 of farm products in 1984" (Leistritz and others 1989, 1). An organization of organic farmers from the Northern Plains region made available their membership list for the supplementary sample.

Table A-2. Response rates for Phase I.

State (Total Farms)	Eligible: Farm Sample + Supplementary Sample = Total	Interviewed by phone: Farm Sample + Supplementary Sample = Total	Returned Questionnaire: Farm Sample + Supplementary Sample = Total
Iowa (97,000)	1,204 + 177 = 1381	1,067 + 169 = 1,236	764 + 141 = 905
Minnesota (88,000)	800 + 653 = 1,453	—	504 + 512 = 1,016
Montana (24,000)	1,100 + 84 = 1,184	—	539 + 59 = 598
North Dakota (33,000)	463 + 71 = 534	340 + 56 = 396	230 + 46 = 276
Totals	**3,567 + 985 = 4552**	**1,407 + 225 = 1,632**	**2,037 + 758 = 2,795**

Operational Definitions Developed for Sustainable Agriculture

Each state developed its own sustainability measure to fit its unique agricultural patterns. However, the similarity of these measures among the four states is remarkable (Chapter 3, Table 3-1), especially given the varied agricultural and ecological settings from which respondents were drawn. The criteria in Table 3-1 describe sustainability within each state, but we had to go down another level because important *regional* differences also exist within each state. Each has three or more distinct subregions with different mixes of farm enterprises.

Iowa. Iowa respondents were sampled from five agricultural subregions: "western livestock," "northeastern dairy and grain," "central cash grain," "southern pasture, cow-calf," and "southeastern mixed crops." To reflect regional differences in Iowa's agricultural systems, the sustainability measures were *normed.* This means that they were assigned numerical values based on the position of a response relative to all other responses from that same subregion.

The practical effect of norming is to guarantee that at least some farmers from all regions and production systems fall into the sustainable and conventional categories. This is so because the reference point for this determination is the production systems of each region, which may be distinct from those used elsewhere.

Minnesota. In Minnesota, we examined differences in the sustainable practices index across seven regions. However, specific sustainability measures were not normed to adjust for regional locations of respondents. Therefore, dairy farms from one region were compared to sugar beet farms from another region in defining conventional and sustainable farms. This should be borne in mind when evaluating the Minnesota data in Chapters 4 through 9.

Montana. Being a state with mountain resources as well as extensive public lands and forests, Montana is less easily divided into contiguous regions. However, we distinguished three broad types of operations: dryland farms, extensive livestock ranches, and irrigated crop-livestock operations. The sustainability index was applied to each type. Table A-3 presents the *collective* indicators used for all three types; the sustainable agriculture index for each type of operation used a subset of these indicators (that is, it was normed).

North Dakota. North Dakota was divided into three regions—western, central, and eastern—based on soil type, climate, and cropping patterns. Except for value orientations, responses to each of the remaining components of the sustainability index were normed to reflect regional differences.

Phase II: In-Depth Interviews

Table A-2 shows interview response rates in each state. These reflect the different approaches used to obtain respondents. Minnesota and Montana used state agricultural agencies for selecting farm samples, and personal interviews required permission from respondents to allow the agencies to release their names. This brought a low response in Minnesota, so its sustainable sample is less representative than desired. As a result, some farms classified as "sustainable" actually fell closer to the center of the sustainability index than was true for other states. Thus, distinctions drawn in later chapters between conventional and sustainable farms in Minnesota are less attributable to their differences in farming practices.

Table A-3. Response rates for Phase II

State	Conventional Farmers: Responded/Targeted: Percent	Sustainable Farmers: Responded/Targeted : Percent
Iowa	52/55: 95%	55/56: 98%
Minnesota	19/60: 32%	22/61: 36%
Montana	24/100: 24%	28/60: 47%
North Dakota	38/60: 63%	41/60: 68%

Obtaining permission for the Phase II interviews was more straightforward for Montana's supplementary sample and for all respondents from Iowa and North Dakota, where researchers directly selected participants and obtained their permission to interview. However, the extent to which the sustainable samples were composed of members of sustainable farming organizations differed by state.

Implications of Methodological Differences for Research Findings

Despite an effort to maintain uniformity in our research methods across the four states, some variation exists. This resulted from the different ways in which each state's research efforts were organized, the types of alternative farming organizations sampled in each state, and the highly diverse agricultural production systems.

Although membership in alternative farming organizations was not a formal criterion for classification as "sustainable," membership lists of farming organizations were used to develop supplemental samples in two states: one sustainable organization in North Dakota and several sustainable agricultural organizations in Iowa. This sampling of organizational members contrasted with our use in Minnesota and Montana of mailing lists of persons thought to be interested in sustainable agriculture.

The following cautions should be borne in mind when comparing results across the four states:

1. *Evidence suggests that organizational membership can contribute significantly to the success of sustainable farmers.* Persons attempting to implement the new farming practices without the benefit of an organizational knowledge base are at a distinct disadvantage.

2. *Important differences exist between organic agriculture and sustainable agriculture,* yet some view them as part of the same alternative farming movement. Organic farming organizations have been around longer, and have a more experienced membership, certification procedures, and written guidelines. This difference may help explain some North Dakota patterns, where an organic farming organization was the source of the supplementary sample for Phase I.

3. *Minnesota did not norm their sustainability indices within regions, or for similar types of producers.* Minnesota used a single index for all respondents, regardless of their specific farming system or region. Relying upon a single index can produce a greater concentration of farmers from a limited number of regions or production systems than is achieved with a norming procedure. Consequently, distinctions between conventional and sustainable farms in Minnesota may reflect regional differences as well as differences in farming practices.

APPENDIX B
ORGANIZATIONS AND GRANTS

Grants and their numbers are shown with date awarded.

Alternative Energy Resources Organization, Helena, Montana

To work with Montana State University to plan on-farm research on sustainable agriculture, as part of the Foundation's proposed multistate analysis of the impacts of sustainable agriculture. *(88-74) 12/12/88*

To conduct a comprehensive assessment of the social, economic, and environmental aspects of sustainable agriculture in Montana, as part of the Foundation's multistate analysis of the impacts of sustainable agriculture. *(89-60) 10/13/89*

To ensure the continued involvement of Alternative Energy Resources Organization in the Foundation's multistate initiative examining the environmental, economic, and social implications of conventional and sustainable agriculture. (92-64) *11/13/92*

To prepare an annotated mailing list of key policymakers; media; and farm, consumer, and environmental leaders for use by the Foundation and others in the dissemination of research results of work on sustainable agriculture. *(93-94) 10/12/93*

The Center for Rural Affairs, Walthill, Nebraska

To coordinate an effort to bring together all of the major locally controlled sustainable agriculture groups in the Foundation's region. *(88-29) 7/25/88*

To support the organizations participating in the Foundation's initiative in sustainable agriculture. *(88-61) 12/09/88*

To coordinate the completion of the examination of the economic, social, and environmental impacts of both conventional and sustainable agriculture, and to assist in the dissemination of results from all the Foundation's sustainable agriculture grants. *(91-65) 11/08/91*

Henry A. Wallace Institute for Alternative Agriculture, Greenbelt, Maryland

To ensure the utilization of recent Foundation-supported findings that focus on establishing new standards and practices for agricultural research at land-grant universities. *(93-96) 11/05/93*

Iowa Coalition of Community Organizations, Des Moines, Iowa

To continue public education and monitoring efforts on groundwater pollution caused by agricultural chemicals. (88-51) 10/14/88

To educate financial institution policymakers on basic sustainable agriculture practices used by farmers. *(92-10) 4/16/92*

Iowa State University, Ames, Iowa, with Practical Farmers of Iowa

To examine the potential economic, social, and environmental impacts on the farm, farm family, and agriculture-dependent rural community of a conversion to low-input sustainable agriculture production practices, as part of the Foundation's multistate analysis of the impacts of sustainable agriculture. *(89-27) 6/08/89*

Izaak Walton League of America, Arlington, Virginia

To assist the Foundation in disseminating the final results of its Sustainable Agriculture Initiative. *11/12/93*

The Land Stewardship Project, Marine, Minnesota

To work with the University of Minnesota to plan a program designed to examine the economic, environmental, and social efficacy of selected sustainable agricultural practices in Minnesota, as part of the Foundation's proposed multistate analysis of the impacts of sustainable agriculture. (88-75) 12/12/88

The Land Stewardship Project, Marine, Minnesota, in cooperation with University of Minnesota Institute of Agriculture, Forestry, and Home Economics

To examine the potential economic, social, and environmental impact on the farm, farm family, and agriculture-dependent rural community of a conversion to low-input sustainable agriculture production practices, as part of the Foundation's multistate analysis of the impacts of sustainable agriculture. *(89-28) 6/08/89*

To cover the indirect costs of the Land Stewardship Project's participation in the Foundation's multistate analysis of the economic, social, and environmental consequences of a transition to sustainable agriculture. *(89-115) 2/24/90*

For the indirect costs of participation in the Foundation's multistate analysis of the economic, social, and environmental consequences of a transition to sustainable agriculture. *(91-61) 9/18/91*

To ensure the Land Stewardship Project's continued involvement in the Foundation's multistate initiative examining the environmental, economic, and social implications of conventional and sustainable agriculture. *(92-56) 9/11/92*

Minnesota Food Association, St. Paul, Minnesota

To explore the creation of an independent institute focusing on sustainable agriculture research and education in the Upper Midwest. (88-85) 2/03/89

University of Minnesota Center for International Agricultural Policy, St. Paul, Minnesota

For public policy research and education on the potential impact in the Foundation's region of international environmental regulations on agriculture. (90-16) 4/12/90

University of Nebraska/Lincoln, Lincoln, Nebraska

To ensure the participation of farmers, alternative agriculture groups, and research institutions from within the Foundation's region in the establishment of the North Central School for Sustainable Agricultural Systems. (93-68) 7/6/93

North Dakota State University in cooperation with Northern Plains Sustainable Agriculture Society, Carrington, North Dakota

To support planning with the NPSAS for a collaborative project to examine the efficacy of sustainable agricultural practices in North Dakota, as part of the Foundation's proposed multistate analysis of the impacts of sustainable agriculture. (88-80) 1/03/89

To examine the potential economic, social, and environmental impact on the farm, farm family, and agriculture-dependent rural community

263

of a conversion to low-input sustainable agriculture production practices, as part of the Foundation's multistate analysis of the impacts of sustainable agriculture. (89-34) 6/08/89

To expand the University's collection and analysis of economic and sociological data concerning agricultural practices on conventional, sustainable, and transitional farms in North Dakota. (91-76) 12/05/91

North Dakota State University, Carrington Research Center, Carrington, North Dakota

To complete the development, field testing, and utilization of a set of soil quality indicators that could help develop policy to improve the economic and environmental performance of American agriculture. (92-35) 7/09/92

Northern Plains Sustainable Agriculture Stewardship Fund, Maida, North Dakota

To ensure the continued involvement of Northern Plains Sustainable Agriculture Society in the Foundation's multistate initiative examining the environmental, economic, and social implications of conventional and sustainable agriculture. (93-1) 3/15/93

Oregon State University with Oregon Tilth, Corvallis, Oregon

To plan a series of on-farm demonstrations designed to explore the potential of sustainable agriculture, as part of the Foundation's multistate analysis of the impacts of sustainable agriculture. (88-76) 12/12/88

To examine the economic, social, and environmental consequences of a conversion to sustainable agriculture practices in the maritime regions of the Pacific Northwest, as part of the Foundation's multistate analysis of the impacts of sustainable agriculture. (89-71) 10/13/89

Oregon State University Department of Crop and Soil Science, Corvallis, Oregon

To underwrite the development, production, and distribution of a guide to sustainable agriculture resources in the Pacific Northwest. (92-17) 5/14/92

APPENDIX B

Oregon Tilth, Tualatin, Oregon

To examine the economic, social, and environmental consequences of a conversion to sustainable agriculture practices in the maritime regions of the Pacific Northwest, as part of the Foundation's multistate analysis of the impacts of sustainable agriculture. (89-71A) 10/13/89

To expand Oregon Tilth's outreach and dissemination efforts on alternative low-input food systems into eastern Oregon and Washington and Idaho. (91-10) 4/05/91

To ensure the continued involvement of Oregon Tilth in the Foundation's multistate initiative examining the environmental, economic, and social implications of conventional and sustainable agriculture. (92-67) 11/13/92

Practical Farmers of Iowa with Iowa State University, Ames, Iowa

To work with Iowa State University to plan on-farm research on the social, economic, and biological aspects of sustainable agriculture in Iowa, as part of the Foundation's proposed multistate analysis of the impacts of sustainable agriculture. (88-77) 12/12/88

To ensure the continued involvement of Practical Farmers of Iowa in the Foundation's multistate initiative examining the environmental, economic, and social implications of conventional and sustainable agriculture. (93-2) 3/15/93

Soil Conservation Society of America, Ankeny, Iowa

For a systematic examination of how farmers and ranchers are persuaded to change their agricultural practices so as to reduce negative environmental effects. (92-49) 9/11/92

South Dakota State University Economics and Plant Science Departments, Brookings, South Dakota

To conduct research on the economic viability of sustainable agriculture practices and the effect of current federal policies on the adoption of such practices. (88-56) 10/14/88

South Dakota State University Economics Department, Brookings, South Dakota, in cooperation with Northern Plains Sustainable Agriculture Society

To support planning with the Northern Plains Sustainable Agriculture

Society for a collaborative project to examine the efficacy of sustainable agricultural practices in South Dakota as part of the Foundation's multistate analysis of the impacts of sustainable agriculture. (88-60) 1/03/89

Tufts University School of Nutrition, Medford, Massachusetts
To develop new standards and practices of agricultural research for land-grant universities, to ensure the unbiased examination of sustainable agriculture practices. (90-51) 6/14/90

Virginia Polytechnic Institute and State University, Blacksburg, Virginia
To examine the impact of a conversion to low-input sustainable agricultural practices on agriculture-dependent rural communities in the Foundation's region, as part of the Foundation's multistate analysis of the impacts of sustainable agriculture. (89-42) 6/08/89

Washington State University Department of Agricultural Economics, Pullman, Washington
For multidisciplinary research on the environmental and economic impacts of current federal commodity policies in the Palouse region of Washington, Oregon, and Idaho. (88-73) 12/09/88

Wisconsin Rural Development Center, Inc., Mount Horeb, Wisconsin
To evaluate the Foundation's approximately $4 million Sustainable Agriculture Initiative. (93-13) 3/12/93

University of Wisconsin/La Crosse, La Crosse, Wisconsin
To prepare a series of maps with appropriate text to illustrate the diversity of geophysical conditions and patterns affecting agriculture within the Foundation's region. (93-90) 9/10/93

University of Wisconsin/La Crosse, La Crosse, Wisconsin
To provide supplemental funds to cover the cost of maps requested by Foundation staff. (94-33) 6/6/94

REFERENCES

Allen, Timothy F. H., and Thomas W. Hoekstra. 1992. *Toward a unified ecology.* New York: Columbia University Press. [Chapter 12]

Allen, Patricia, and Carolyn Sachs. 1993. Sustainable agriculture in the United States: Engagements, conditions and contradictions of sustainability. In *Food for the Future,* ed. Patricia Allen. New York: John Wiley & Sons. [Chapter 3]

Allen, Patricia, Debra Van Dusen, Jackelyn Lundy, and Stephen Gliessman. 1991. Integrating social, environmental, and economic issues in sustainable agriculture. *American Journal of Alternative Agriculture* 6(1): 34-39. [Chapter 3]

Barnes, Donna, and Audie Blevins. 1992. Farm structure and the economic well-being of non-metropolitan counties. *Rural Sociology* 57(3): 333-46. [Chapter 7]

Batte, Marvin T. 1992. Sustainable agriculture and farm profitability: The case of organic farming in Ohio. Presented at American Agricultural Economics Association Annual Meeting, Baltimore, MD. [Sidebar 6-1]

Berry, Wendell. 1977. *The unsettling of America: Culture and agriculture.* San Francisco: Sierra Club Books. [Part II Intro]

Beus, Curtis, and Riley Dunlap. 1990. Conventional versus alternative agriculture: The paradigmatic roots of the debate. *Rural Sociology* 55(4): 590-616. [Chapter 4] [Sidebar 9-1]

Bird, Alan. 1992. Virtual large farms and exurban communities: Keys to sustainable agriculture. *Choices* (third quarter): 54-55. [Chapter 4]

Bird, Elizabeth. 1994. Reviewing commitments to sustainable agriculture research. *Consortium News* 1:3 (February). Published by Center for Rural Affairs, Walthill, Nebraska. [Chapter 16]

Bird, Elizabeth. 1991. *Research for sustainability: The National Research Initiative's social plan for agriculture.* Walthill, Nebraska: Center for Rural Affairs. [Chapter 16]

Bird, Elizabeth, and Chuck Hassebrook. 1992. *Report card on USDA research policy.* Walthill, Nebraska: Center for Rural Affairs Special Report, November. [Chapter 16]

Blackmer, A. M., T. F. Morris, and G. D. Binford. 1992. Predicting N [nitrogen] fertilizer needs for corn in humid regions: Advances in Iowa. In *Predicting N fertilizer needs in humid regions,* ed. B. R. Bock and K. R. Kelley. Bulletin Y226. National Fertilizer and Environmental Research Center, Tennessee Valley Authority, Muscle Shoals, AL 35660. [Sidebar 7-2]

Borchert, John R., and Russel B. Adams. 1963. *Trade centers and trade areas of the upper Midwest.* Minneapolis: Upper Midwest Council, Urban Report No. 3. [Chapter 7]

Bultena, Gordon L., and Eric O. Hoiberg. 1992. Farmers' perceptions of the personal benefits and costs of adopting sustainable agricultural practices. *Impact Assessment Bulletin* 10(2): 43-57. [Sidebar 3-2] [Sidebar 3-5]

Bultena, Gordon L., Eric O. Hoiberg, Susan K. Jarnagin, and Rick Exner. 1992. *Transition to a more sustainable agriculture in Iowa: A comparison of the orientations and farming practices of conventional, transitional, and sustainable farm operators.* Ames, Iowa: Iowa State University, Department of Sociology, Sociology Report 166. [Chapter 4] [Chapter 5] [Chapter 15] [Sidebar 6-2] [Appendix A]

Bultena, Gordon, and Larry Leistritz, eds. 1990. Projected impacts from sustainable agriculture. *Impact Assessment Bulletin* 10(2), special issue. [Part II Intro]

Bureau of the Census, U.S. Department of Commerce. 1990a. *Census of agriculture, 1987: Agricultural atlas of the United States.* Washington. [Chapter 2]

Bureau of the Census, U.S. Department of Commerce. 1990b. *Census of agriculture, 1987,*

REFERENCES

on CD-ROM: County data. Washington. [Chapter 2]

Bureau of the Census, U.S. Department of Commerce. 1990. *Census of population, 1990.* Washington. [Chapter 2]

Cacek, Terry, and Linda L. Langner. 1986. The economic implications of organic farming. *American Journal of Alternative Agriculture* 1(1): 25-29. [Sidebar 6-1]

Carr, P.M., G.R. Carlson, J.S. Jacobsen, G.A. Nielsen, and E.O. Skogley. 1991. Farming soils, not fields: a strategy for increasing fertilizer profitability. *Journal of Production Agriculture* 4: 57-61. [Chapter 3]

Chase, Craig, and Michael Duffy. 1991. An economic comparison of conventional and reduced-chemical farming systems in Iowa. *American Journal of Alternative Agriculture* 6(4): 168-73. [Sidebar 6-1]

Clancy, S.A., J. C. Gardner, C. E. Grygiel, M. E. Biondini, and G. K. Johnson. 1993. *Farming practices for a sustainable agriculture in North Dakota.* Fargo, North Dakota: North Dakota Agricultural Experiment Station Bulletin, available from the North Dakota State University Carrington Research Extension Center. [Chapter 10] [Chapter 12]

Coleman, D. C. 1989. Ecology, agroecosystems, and sustainable agriculture. *Ecology* 70: 1590. [Chapter 12]

Cordray, Sheila M., and Kathryn W. Goetz. 1991. *Issues in sustainable agriculture: A study of horticultural producers in western Oregon and Washington.* Corvallis, Oregon: Department of Sociology, Oregon State University. [Appendix A]

Cordray, Sheila M., Larry S. Lev, Richard P. Dick, and Helene Murray. 1993. Sustainability of Pacific Northwest horticultural producers. *Journal of Production Agriculture* 6(1): 121-25. [Chapter 15] [Sidebar 3-3]

Crews, Timothy E., Charles L. Mohler, and Alison G. Power. 1991. Energetics and ecosystem integrity: The defining principles of sustainable agriculture. *American Journal of Alternative Agriculture* 6(3): 146-49. [Sidebar 6-1]

Crookston, R. Kent, and Melvin J. Stanford. 1990. *Dick and Sharon Thompson's "Problem Child," a Decision Case in Sustainable Agriculture.* St. Paul, Minnesota: Department of Agronomy and Plant Genetics, University of Minnesota.

Crosson, Pierre, and Janet Ekey Ostrov. 1990. Sorting out the environmental benefits of alternative agriculture. *Journal of Soil and Water Conservation* 45(1): 34-41. [Sidebar 6-1]

Daly, H. E., and J. B. Cobb. 1989. *For the common good.* Boston: Beacon Press. [Chapter 12]

Davidson, Osha Gray. 1990. *Broken heartland: The rise of America's rural ghetto.* New York: Anchor Books. [Chapter 8]

Diebel, Penelope L., Richard V. Llewelyn, and Jeffrey R. Williams. 1993. A yield sensitivity analysis of conventional and alternative whole-farm budgets for a typical northeast Kansas farm. Presented at Western Agricultural Economics Association Annual Meeting, Edmonton, Canada. [Sidebar 6-1]

· *Doane's Agricultural Report,* July 12, 1992. Farm financial ratios. [Chapter 6]

Dobbs, Thomas L. 1994. Organic, conventional, and reduced till farming systems: Profitability in the Northern Great Plains. *Choices* 9(2):31-32. [Sidebar 6-1]

Dobbs, Thomas L. 1993. *Implications of sustainable farming systems in the northern Great Plains for farm profitability and size.* Brookings, South Dakota: South Dakotas State University, Department of Agricultural Economics, Economic Staff Paper 93-5. [Chapter 4]

REFERENCES

Dobbs, Thomas L. 1992. Testimony at hearing on "Agricultural Industrialization and the Family Farm: The Role of Federal Policy," held by Joint Economic Committee of the U.S. Congress, Washington, July 8. [Sidebar 16-1]

Dobbs, Thomas L., and David L. Becker. 1992. Mandatory supply controls versus flexibility policy options for encouraging sustainable farming systems. *American Journal of Alternative Agriculture* 7(3): 122-28. [Sidebar 16-1]

Dobbs, Thomas L., and John D. Cole. 1992. Potential effects on rural economies of conversion to sustainable farming systems. *American Journal of Alternative Agriculture* 7(1/2): 70-80. [Chapter 7]

Dobbs, Thomas L., and Lon D. Henning. 1993. Long-term economic performance of alternative, conventional, and reduced tillage farming systems in east-central South Dakota. Brookings: South Dakota State University, Economic Commentator 330. [Sidebar 6-1]

Dobbs, Thomas L., Donald C. Taylor, and James D. Smolik. 1992. *Farm, rural economy, and policy implications of sustainable agriculture in South Dakota.* Brookings, South Dakota: South Dakota State University Agriculture Experiment Station Bulletin 713. [Sidebar 6-1] [Sidebar 16-1]

Duffield, John, Catherine Berg, Douglas Dalenberg, Kay Unger, and Chris Neher. 1993. *Economic evaluation of sustainable agricultural practices in Montana (final report of the Montana Sustainable Agriculture Assessment Project).* Missoula, Montana: Bioeconomics, Inc. [Sidebar 6-6]

Eberhardt, Bruce J., and Abdullah Pooyan. 1990. Development of the farm stress survey: Factorial structure, reliability, and validity. *Educational and Psychological Measurement* 50(2): 393-402. [Chapter 8]

Elliot, E. T., and C. V. Cole. 1989. A perspective on agroecosystem science. *Ecology* 70: 1597-1602. [Chapter 12]

Exner, Rick. 1990. *On-Farm research: Using the skills of farmers and scientists—Practical Farmers of Iowa and Iowa State University.* In Extending Sustainable Systems, Proceedings of a training conference on sustainable agriculture, St. Cloud, Minnesota, May 9-10, University of Minnesota. [Chapter 10]

Faeth, Paul, Robert Repetto, Kim Kroll, Qi Dai, and Glenn Helmers. 1991. *Paying the Farm Bill: U.S. Agricultural Policy and the Transition to Sustainable Agriculture.* World Resources Institute, Washington, D.C. [Sidebar 6-1]

Fals-Borda, O., and M. A. Rahman. 1991. *Action and knowledge: Breaking the monopoly with participatory action-research.* New York: Apex Press. [Chapter 10]

Fedle, D. M. and Bill Anderson (Agri Control Company, Sioux City, Iowa). 1992. Do the task force recommendations work? *Ag Executive* (July). [Chapter 6]

Fliegel, Frederick C. 1993. *Diffusion research in rural sociology.* Westport, Connecticut: Greenwood Press. [Chapter 9]

Flora, Cornelia Butler. 1994. Sustainable agriculture, sustainable communities: Social capital in the Great Plains and Corn Belt. *Research in Rural Sociology and Development* Volume 6 (annual). [Sidebar 7-1]

Flora, Cornelia Butler. 1990. Sustainability of agriculture and rural communities. In *Sustainable agriculture in temperate zones,* ed. Charles A. Francis, Cornelia Butler Flora, and Larry D. King. New York: John Wiley & Sons, p. 343-60. [Chapter 7] [Chapter 8]

Flora, Cornelia Butler, and Jan Flora. 1988. Public policy, farm size, and community well-being in farming-dependent counties of the Plains. In *Agriculture and community change in the U.S.,* ed. L. E. Swanson. Boulder, CO: Westview Press, p. 76-129. [Chapter 7]

Fox, Glenn, Alfons Weersink, Ghulam Sarwar, Scott Duff, and Bill Deen. 1991. Comparative economics of alternative agricultural production systems: A review.

REFERENCES

Northeastern Journal of Agricultural and Resource Economics (20)1: 124-42. [Chapter 6] [Sidebar 6-1]

Francis, Charles, and Garth Youngberg. 1990. Sustainable agriculture—an overview. In *Sustainable Agriculture in Temperate Zones*, ed. Charles A. Francis, Cornelia Butler Flora, and Larry D. King. New York: John Wiley & Sons. [Chapter 3]

Francis, C., J. King, J. DeWitt, J. Bushnell, and L. Lucas. 1990. Participatory strategies for information exchange. *American Journal of Alternative Agriculture* 5(4): 153-62. [Chapter 10]

Freire, P. 1970. *Pedagogy of the Oppressed*. New York: Continuum Press. [Chapter 10]

Gale, Fred, and David Henderson. 1991. *Estimating entry and exit of U.S. farms*. Washington: U.S. Department of Agriculture, Agriculture and Rural Economy Division, Staff Report AGES 9119. [Chapter 1]

Gerber, John. 1991. "Participatory research and education for agricultural sustainability." In Participatory On-farm Research Concepts and Implications. Urbana, Illinois: University of Illinois Agro-Ecology Program Paper 91-15, October. [Chapter 10]

Gips, Terry, 1984. What is sustainable agriculture? *Manna* July/August. [Chapter 3]

Goldschmidt, Walter. 1978. *As you sow: Three studies in the social consequences of agribusiness*. Montclair, New Jersey: Allanheld, Osmun, & Company. [Sidebar 3-5] [Chapter 7]

Goreham, Gary A., George A. Youngs, Jr., and David L. Watt. 1992. *A comparison of sustainable and conventional farmers in North Dakota: Final report on Phase II of the Northwest Area Foundation Initiative on Sustainable Agriculture*. Fargo, North Dakota: North Dakota State University, Departments of Sociology/Anthropology and Agricultural Economics. [Part II Intro]

Goreham, Gary A., F. Larry Leistritz, and Richard W. Rathge. 1986. *Trade and marketing patterns of North Dakota farm and ranch operators*. Fargo: North Dakota State University, Department of Agricultural Economics, Agricultural Economics Miscellaneous Report No. 98 (September). [Chapter 7]

Goss, Kevin. 1979. Consequences of diffusion of innovations. *Rural Sociology* 44 (Winter): 754-72. [Chapter 9]

Granatstein, David. 1988. *Reshaping the bottom line: On-farm strategies for a sustainable agriculture*. Lewiston, Minnesota: Land Stewardship Project. [Chapter 10]

Green-McGrath, D., L. S. Lev, H. Murray, and R. D. William. 1993. Farmer/scientist focus sessions: A how-to guide. Corvallis, Oregon: Oregon State University Extension Service Bulletin No. EM 8554. [Chapter 10]

Hallberg, Milton. 1992. *Policy for American agriculture: Choices and consequences*. Ames, Iowa: Iowa State University Press. [Chapter 4]

Hanson, James C., Dale M. Johnson, Steven E. Peters, and Rhonda R. Janke. 1990. The profitability of sustainable agriculture on a representative grain farm in the mid-Atlantic region, 1981-89. *Northeastern Journal of Agriculture and Resource Economics* 19(2): 90-98. [Sidebar 6-1]

Haynes, Michael N., and Alan L. Olmstead. 1984. Farm size and community quality: Arvin and Dinuba revisited. *American Journal of Agricultural Economics* 66(4): 430-36. [Chapter 7]

Ikerd, John, 1992. *Sustainable Agriculture and Quality of Life*, A report of the Sustainable Agriculture Quality of Life Task Force, Sustainable Agriculture Research and Education Program, USDA, Washington, DC. [Chapter 16]

Ikerd, John, Sandra Monson, and Donald Van Dyne. 1993. Alternative farming systems for U.S. agriculture: New estimates of profit and environmental effects. *Choices* 8(3):37-38. [Chapter 6] [Sidebar 6-1]

Iowa State University Extension. 1993. *Estimated returns from feeding livestock.* Ames, Iowa: Iowa State University. [Sidebar 6-3]

Jackson, W. 1985. *New roots for agriculture.* Second edition. Lincoln, Nebraska: University of Nebraska Press. [Chapter 4] [Chapter 12]

Jackson, W., and J. Piper. 1989. The necessary marriage between ecology and agriculture. *Ecology* 70: 1591-93. [Chapter 12].

Jamtgaard, Keith. 1992a. *Results from the Montana Agricultural Assessment Project—Phase II interviews with sustainable and conventional producers.* Bozeman, Montana: Department of Sociology, Montana State University. [Part II Intro] [Appendix A]

Jamtgaard, Keith. 1992b. *Results from the Montana Agricultural Assessment Questionnaire: A survey of sustainable agriculture.* Helena, Montana: Alternative Energy Resources Organization (AERO). [Chapter 10]

Jefferson Davis Associates. 1982. *Soil Conservation attitudes and practices: The present and future.* Cedar Rapids, Iowa: Prepared for Pioneer Hi-Bred, Inc. [Chapter 10]

Keeney, Dennis. 1989. Toward a sustainable agriculture: Need for clarification of concepts and terminology. *American Journal of Alternative Agriculture.* 4(3-4): 101-5. [Chapter 3]

Kirschenmann, Frederick. 1992. What can alternative farming systems and rural communities do for each other? In *Alternative Farming Systems and Rural Communities: Exploring the Connections,* 25-38. Proceedings of the ninth annual symposium of the Institute for Alternative Agriculture, Chevy Chase, Maryland. [Part II Intro] [Chapter 8]

Klepper, R., W. Lockeretz, B. Commoner, M. Gertler, S. Fast, D. O'Leary, and R. Blobaum. 1977. Economic performance and energy intensiveness on organic and conventional farms in the corn belt: A preliminary comparison. *American Journal of Agricultural Economics* 59(1): 1-12. [Sidebar 6-2]

Kloppenburg, Jack. 1992. Social theory and the de/reconstruction of agricultural science: Local knowledge for an alternative agriculture. *Rural Sociology* 56(4): 519-48. [Chapter 9]

Kloppenburg, Jack, and Neva Hassanein. 1993. As you know: Farmers' knowledge and sustainable agriculture. *As you sow* (a periodic newsletter). University of Wisconsin/Madison, Department of Rural Sociology, publication 27. [Chapter 9]

Korsching, Peter F. 1985. Farm Structural Characteristics and Proximity of Purchase Location of Goods and Services. P. 261-87 in *Research in rural sociology and development, volume 2,* ed. Frank A. Fear and Harry K. Schwarzweller. Greenwich, Connecticut: Jai Press, Inc. [Chapter 7]

Kramer, Mark. 1980. *Three farms: Making milk, meat, and money from the soil.* Boston: Little, Brown. [Chapter 8]

Kroese, Ron. 1988. *Excellence in agriculture: Interviews with ten Minnesota Stewardship farmers.* Lewiston, Minnesota: Land Stewardship Project. [Chapter 10]

Labao, Linda M. 1990. *Locality and inequality: Farm and industry structure and socioeconomic conditions.* Albany: State University of New York Press. [Sidebar 3-5] [Chapter 7]

Land Stewardship Project. 1994. *An agriculture that makes sense: Profitability of four sustainable farms in Minnesota.* A 12-minute video available from Land Stewardship Project, P. O. Box 130, Lewiston, Minnesota 55952, 507/523-3366. [Sidebar 4-2] [Sidebar 6-5]

Lasley, Paul, Eric Hoiberg, and Gordon Bultena. 1993. Is sustainable agriculture an elixir for rural communities? Community implications of sustainable agriculture. *American*

REFERENCES

Journal of Alternative Agriculture 8(3): 133-39. [Part II Intro] [Chapter 7] [Chapter 8]

Lasley, Paul, Michael Duffy, Kevin Kettner, and Craig Chase. 1990. Factors affecting farmers' use of practices to reduce commercial fertilizers and pesticides. *Journal of Soil and Water Conservation* 45(1): 132-36. [Chapter 4] [Chapter 15]

Leistritz, F. Larry, Brenda L. Ekstrom, Janet Wanzek, and Timothy L. Mortensen. 1989. *Outlook of North Dakota farm households: Results of the 1988 longitudinal farm survey.* Fargo, North Dakota: Agricultural Economics Report No. 246, Department of Agricultural Economics, North Dakota State University. [Appendix A]

Liebhardt, W. C., R. W. Andrews, M. N. Culik, R. R. Harwood, R. R. Janke, J. K. Radke, and S. L. Rieger-Schwartz. 1989. Crop production during conversion from conventional to low-input methods. *Agronomy Journal* 81(2): 150-59. [Sidebar 6-2]

Lockeretz, William. 1988. Open questions in sustainable agriculture, *American Journal of Alternative Agriculture* 3(4): 174-81. [Chapter 3]

Lockeretz, William. 1986. Alternative Agriculture.: In *New directions for agriculture and agricultural research: Neglected dimensions and emerging alternatives,* ed. Kenneth A. Dahlberg. Totowa, New Jersey: Rowman & Allanheld, p. 291-311. [Chapter 7]

Lockeretz, William, and Molly D. Anderson. 1993. *Agricultural Research Alternatives.* Lincoln, Nebraska: University of Nebraska Press. [Chapter 16]

Lockeretz, William, Georgia Shearer, and Daniel H. Kohl. 1981. Organic farming in the corn belt. *Science* 211(6): 540-46. [Sidebar 6-1]

Loucks, Orie L. 1970. Evolution of diversity, efficiency and community stability. *American Zoologist* 10:17-25. [Chapter 12]

MacCannell. Dean. 1983. "Agribusiness and the small community." Washington: U.S. Congress, Office of Technology Assessment, background paper to Technology, Public Policy, and the Changing Structure of American Agriculture. [Chapter 1]

MacRae, Rod, Stuart Hill, John Henning, and Alison Bentley. 1990. Policies, programs, and regulations to support the transition to sustainable agriculture in Canada. *American Journal of Alternative Agriculture* 5(2): 76-92. [Sidebar 3-5]

Madden, J. Patrick, and Thomas L. Dobbs. 1990. The role of economics in achieving low-input farming systems. Ch. 26 in *Sustainable Agricultural Systems,* ed. Clyve A. Edwards et al., Ankeny, Iowa: Soil and Water Conservation Society, 459-77. [Chapter 6] [Sidebar 6-1]

Mann, Susan. 1990. *Agrarian capitalism in theory and practice.* Chapel Hill, North Carolina: University of North Carolina Press. [Chapter 5]

Mansson, B. A., and J. M. McGlade. 1993. Ecology, thermodynamics, and H. T. Odum's conjectures. *Oecologia* 93: 582-96. [Chapter 12]

Matheson, Nancy. 1993. AERO Farm Improvement Clubs: A collaborative learning community. *Journal of Pesticide Reform* (Spring), Eugene, Oregon. Also contact: Alternative Energy Resources Organization (AERO), 25 S. Ewing, Suite 214, Helena, MT 59601 (406/443-7272; Fax: 406/442-9120).

Matheson, Nancy (ed.). 1989. *Sustainable agriculture in the Northern Rockies and Plains.* Helena, Montana: Alternative Energy Resources Organization. [Chapter 10] [Sidebar 10-1] [Chapter 11]

Matheson, Nancy, Barbara Rusmore, James R. Sims, Michael Spengler, and E. L. Michalson. 1991. *Cereal-legume cropping systems: Nine farm case studies in the dryland Northern Plains, Canadian prairies, and Intermountain Northwest.* Helena, Montana: Alternative Energy Resources Organization. ($6.00 postpaid in U.S.; quantity discount available; AERO, 44 North Last Chance Gulch #9, Helena, MT 59601; 406/443-7272)

272

REFERENCES

[Chapter 11]

Menanteau-Horta, Dario, Virginia M. Juffer, and Bruce Maxwell. 1993. *Patterns and trends of sustainable agriculture: A comparison of selected Minnesota farmers.* St. Paul: Center for Rural Social Development, University of Minnesota. [Part II Intro]

Menanteau-Horta, Dario, Virginia M. Juffer, Bruce Maxwell, and Ron Kroese. 1991. *Sustainable agriculture in Minnesota.* St. Paul: Center for Rural Social Development, University of Minnesota. [Appendix A]

Murray, Helene, Larry S. Lev, Daniel Green-McGrath, and Alice Mills Morrow. 1994a. *Whole-farm case studies: A how-to manual.* Oregon State University Extension Service EM 8558, 8 p. [Chapter 11]

Murray, Helene, Richard Dick, Daniel Green-McGrath, Lorna Michael Butler, Larry S. Lev, and Richard Carkner. 1994b. *Whole farm case studies of horticultural crop producers in the maritime Pacific Northwest.* Oregon State University Station Bulletin 678, 28 p. [Chapter 11] [Sidebar 11-1]

National Research Council. 1989. *Alternative agriculture.* Washington: National Academy Press. [Chapter 1] [Chapter 6] [Chapter 9] [Chapter 15] [Part II Intro]

National Sustainable Agriculture Coordinating Council (NSACC). 1994. *The campaign for sustainable agriculture: Working toward a new direction in federal farm and food policy.* Available from NSACC, 32 North Church St., Goshen, NY 10924. [Chapter 3]

Nielsen, E., and L. Lee. 1987. *The magnitude and costs of groundwater contamination from agricultural chemicals.* Washington: U.S. Department of Agriculture Economic Research Service, Agricultural Economic Report No. 576, October. [Chapter 1]

Norris, Patricia E., and Leonard A. Shabman. 1992. Economic and environmental considerations for nitrogen management in the mid-Atlantic coastal plain. *American Journal of Alternative Agriculture* 7(4): 148-56. [Sidebar 6-1]

Nowak, Peter J. 1983. Adoption and diffusion of soil and water conservation practices. *The Rural Sociologist* 3 (March): 83-91. [Chapter 9]

Odum, E. P. 1969. The strategy of ecosystem development. *Science* 164: 262-70. [Chapter 12]

Odum, H. T. 1983. *Systems ecology.* New York: Wiley. [Chapter 12]

Olson, Kent D., and Douglas Tvedt. 1987. *On comparing farm management associations and the farm population.* St. Paul: University of Minnesota Institute of Agriculture, Forestry, and Home Economics, Department of Agriculture and Applied Economics Staff Paper, 87-29. [Sidebar 6-5]

Olson, K. D., J. Langley, and E. O. Heady. 1982. Widespread adoption of organic farming practices: Estimated impacts on U.S. agriculture. *Journal of Soil and Water Conservation* 37(1): 41-45. [Sidebar 6-2]

Osteen, Craig D., and Philip I. Szmedra. 1989. *Agricultural pesticide use trends and policy issues.* Economic Research Service, U.S. Department of Agriculture Agricultural Economic Report 622. [Chapter 2]

Painter, Kathleen M. 1992. Projecting farm level economic and environmental impacts of farm policy proposals: An interregional comparison. Ph.D. dissertation, Department of Agricultural Economics, Washington State University, Pullman, WA. [Sidebar 16-1]

Painter, Kathleen M., and Douglas L. Young. 1993. Environmental and economic impacts of agricultural policy reform: an interregional comparison. SARE-ACE Project Working Paper No. 93-1, Department of Agricultural Economics, Washington State University, Pullman, WA. [Sidebar 16-1]

Pampel, Fred, and J. C. van Es. 1977. Environmental quality and issues of adoption

REFERENCES

research. *Rural Sociology* 42 (Spring): 57-71. [Chapter 9]

Patten, B. C. 1993. Toward a more holistic ecology and science: The contribution of H. T. Odum. *Oecologia* 93: 597-602. [Chapter 12]

Paul, E. A., and G. P. Robertson. 1989. Ecology and the agricultural sciences: A false dichotomy? *Ecology* 70: 1594-97. [Chapter 12]

Pfeffer, M. J. 1983. Social origins of three systems of farm production in the United States. *Rural Sociology* 48: 540-62. [Chapter 5]

Pugliese, Enrico. 1991. "Agriculture and the New Division of Labor," in *Towards a New Political Economy of Agriculture,* ed. William Friedland, Lawrence Busch, Frederick Buttel, and Alan Rudy. Boulder, Colorado: Westview Press. [Chapter 5]

Rhoades, R. and R. Booth. 1992. "Farmer-back-to-Farmer: Ten Years Later" in Participatory on-farm Research and Education for Agricultural sustainability, Conference proceedings. Urbana, Illinois: Agriculture Experiment Station. [Chapter 10]

Rogers, Everett. 1983. *Diffusion of innovations.* 3rd ed. New York: Free Press. [Chapter 9]

Rural Sociological Society. 1993. *Persistent rural poverty.* Boulder, Colorado: Westview Press. [Chapter 8]

Rusmore, B. 1989. Sustainable agriculture in the Northern Plains and Rocky Mountains. In *Sustainable agriculture in the Northern Rockies and Plains,* ed. N. Matheson. Helena, Montana: Alternative Energy Resources Organization (AERO). [Sidebar 9-2]

Sahs, W. W., G. A. Helmers, and M. R. Langmeier. 1988. Comparative profitability of organic and conventional crop production systems in east-central Nebraska. P. 397-405 in Global perspectives on agroecology and sustainable agricultural systems, ed. P. Allen and D. Van Dusen. Proceedings of the Sixth International Scientific Conference of the International Federation of Organic Agriculture Movements (IFOAM). Vol. 1. Santa Cruz, California: University of California. [Sidebar 6-2]

Skees, Jerry, and Louis Swanson. 1988. Farm structure and local society well-being in the South. In *The rural south in crisis,* ed. L. Beaulieu. Boulder, Colorado: Westview Press, p. 141-57. [Chapter 7]

Smith, M. G. 1988. Older farmers, bigger farms in the offing. *Rural Development Perspectives,* U.S. Department of Agriculture. [Chapter 1]

Smith, Stewart. 1992. "Farming"—It's Declining in the U.S. *Choices* (first quarter) 7:8-10. [Chapter 1]

Smolik, J. D., and T. L. Dobbs. 1991. Crop yields and economic returns accompanying the transition to alternative farming systems. *Journal of Production Agriculture* 4(2): 153-161. [Sidebar 6-2]

Soule, J. D., and J. K. Piper. 1992. *Farming in nature's image.* Washington: Island Press. [Chapter 12]

Stanford, Melvin J., R. Kent Crookston, David W. Davis, and Steve R. Simmons. 1992. *Decision cases for agriculture.* St. Paul: University of Minnesota College of Agriculture Programs for Decision Cases, Library of Congress Catalog Number 92-74432, $22.00 postpaid in U.S. [Chapter 11]

Stearns, L. D., B. L. Dahl, G. A. Goreham, R. M. Jacobsen, R. S. Sell, D. L. Watt, and G. A. Youngs, Jr. 1991. *Selected practice and the financial indicators of sustainable farms in North Dakota.* Fargo, North Dakota: North Dakota State University Department of Agricultural Economics Miscellaneous Report No. 134. [Sidebar 6-2]

Strange, Marty. 1990. *Rural economics development and sustainable agriculture.* Walthill, Nebraska: Center for Rural Affairs. [Chapter 9]

REFERENCES

Strange, Marty. 1988. *Family farming: A new economic vision*. Lincoln: University of Nebraska Press. [Chapter 1] [Part II Intro]

Strange, Marty, Patricia E. Funk, Gerald Hansen, Jennifer Tully, and Don Macke. 1990. *Half a glass of water: State economic development policies and the small agricultural communities of the Middle Border*. Walthill, Nebraska: Center for Rural Affairs. [Chapter 1]

Swanson, Louis. 1990. Rethinking assumptions about farm and community. *In* Luloff, A. E., and L. E. Swanson, eds. *American rural communities*, p. 19-33. Boulder, Colorado: Westview Press. [Chapter 16]

Tennessee Valley Authority. 1982. *Fertilizer summary data*. [Chapter 2]

Tennessee Valley Authority. 1991. *Commercial fertilizers*. [Chapter 2]

Urban, Thomas N. 1991. Agricultural Industrialization: It's Inevitable. *Choices* (fourth quarter): 4-6. [Chapter 1]

U.S. Congress, Office of Technology Assessment. 1986. *Technology, public policy, and the changing structure of American agriculture*. Washington: U.S. Government Printing Office, March, publication OTA-F-285. [Chapter 1] [Chapter 4]

U.S. Council on Environmental Quality. 1989. *Environmental trends*. Washington: U.S. Government Printing Office. [Chapter 2]

U. S. Department of Agriculture. 1981. *Soil and Water Resource Conservation Act: 1980 appraisal*. Washington. [Chapter 2]

U.S. Department of Agriculture, Economic Research Service. 1993. *Economic Indicators of the Farm Sector: National Financial Summary, 1991*. Washington. [Chapter 1]

U.S. Department of Agriculture, Economic Research Service, Agricultural and Rural Economy Division. 1992. *Economic indicators of the farm sector: State financial summary, 1991*. Washington: ECIFS 11-2. [Table 6-1]

U.S. Department of Agriculture, Soil Conservation Service. 1975. *Soil taxonomy*. Agriculture Handbook 436. Washington. [Chapter 2]

U.S. Department of Agriculture, Soil Conservation Service. 1981. *Land resource regions and major land resource areas of the United States*. Agriculture Handbook 296. Washington. (Revision of Agriculture Handbook 296 published in 1965.) [Chapter 2]

U.S. Department of Agriculture, Soil Conservation Service. 1982 National Resources Inventory. Washington. [Chapter 2]

U.S. Department of Agriculture, Soil Conservation Service. 1987 National Resources Inventory. Washington. [Chapter 2]

U. S. Department of Agriculture, Statistical Reporting Service. 1976 (1972). *Commercial fertilizers*. Statistical Bulletin 472, revised. [Chapter 2]

U.S. Geological Survey. 1970. *National atlas of the United States of America*. Reston, VA. [Chapter 2]

Vogel, K. P. 1992. The wrong goal. The world & I. *The Washington Times* 7(2): 252-59. [Chapter 12]

Walker, Lilly S., and James L. Walker. 1987. Stressors and symptoms predictive of distress in farmers. *Family Relations* 36(October): 374-78. [Chapter 8]

Watt, David L., Bruce L. Dahl, Gary A. Goreham, George A. Youngs, Jr., Randall S. Sell, and Larry D. Stearns. 1992. *Characteristics of farms using sustainable practices: North Dakota, 1989, final report, Phase I for the Northwest Area Foundation Initiative*. Fargo, North Dakota: Department of Agricultural Economics, North Dakota State University. [Appendix A]

Wellman, Allen C. 1993. *Crop and livestock prices for Nebraska producers*. Lincoln,

Nebraska: University of Nebraska, Cooperative Extension Service EC93-883-C, June. [Sidebar 6-3]

Westcott, Malvern. 1994. Positive research findings could give a significant marketing advantage to sustainable farmers. Top-quality grain, good yields follow green manure crops. *Sustainable Farming Quarterly* 5(3): 1, 4-6. [Chapter 15]

Youngs, George A., Jr., Gary A. Goreham, and David L. Watt. 1991. Classifying conventional and sustainable farmers: Does it matter how you measure? *Journal of Sustainable Agriculture* 2(2): 91-115. [Chapter 15]